# *SCHAUM'S*
## EASY OUTLINES

# *Biology*

# Online Diagnostic Test

Go to **Schaums.com** to launch the Schaum's Diagnostic Test.

This convenient application provides a 30-question multiple-choice test that will pinpoint areas of strength and weakness to help you focus your study. Questions cover all aspects of biology, and the correct answers are explained in full. With a question-bank that rotates daily, the Schaum's Online Test also allows you to check your progress and readiness for final exams.

## Other titles featured in Schaum's Online Diagnostic Test:

Schaum's Easy Outlines: Calculus, 2nd Edition
Schaum's Easy Outlines: Geometry, 2nd Edition
Schaum's Easy Outlines: Statistics, 2nd Edition
Schaum's Easy Outlines: Elementary Algebra, 2nd Edition
Schaum's Easy Outlines: College Algebra, 2nd Edition
Schaum's Easy Outlines: Human Anatomy and Physiology, 2nd Edition
Schaum's Easy Outlines: Beginning Chemistry, 2nd Edition
Schaum's Easy Outlines: College Chemistry, 2nd Edition
Schaum's Easy Outlines: Organic Chemistry, 2nd Edition

# *Biology*

—————————————— *Second Edition*

*George H. Fried*
*and George J. Hademenos*

*Abridgement Editor:*
*Katherine E. Cullen, Ph.D.*

New York   Chicago   San Francisco   Lisbon   London   Madrid   Mexico City
Milan   New Delhi   San Juan   Seoul   Singapore   Sydney   Toronto

The *McGraw·Hill* Companies

Copyright © 2011 by The McGraw-Hill Companies, Inc. All rights reserved. Printed in the United States of America. Except as permitted under the United States Copyright Act of 1976, no part of this publication may be reproduced or distributed in any form or by any means, or stored in a database or retrieval system, without the prior written permission of the publisher.

2 3 4 5 6 7 8 9 10 11 12 13 14 15     QFR/QFR     1 9 8 7 6 5 4 3 2 1

ISBN   978-0-07-174654-0
MHID      0-07-174654-4

**Library of Congress Cataloging-in-Publication Data**

Fried, George, 1926-
        Schaum's easy outline of biology / George H. Fried. — 2nd ed.
            p.     cm. — (Schaum's outlines)
        ISBN 0-07-174654-4 (alk. paper)
        1. Biology—Outlines, syllabi, etc.     I. Title.

        QH315.5.F73     2009
        570—dc22                                                    2010010887

Trademarks: McGraw-Hill, the McGraw-Hill Publishing logo, Schaum's, and related trade dress are trademarks or registered trademarks of The McGraw-Hill Companies and/or its affiliates in the United States and other countries and may not be used without written permission. All other trademarks are the property of their respective owners. The McGraw-Hill Companies is not associated with any product or vendor mentioned in this book.

McGraw-Hill books are available at special quantity discounts to use as premiums and sales promotions or for use in corporate training programs. To contact a representative, please e-mail us at bulksales@mcgraw-hill.com.

This book is printed on acid-free paper.

# Contents

# Chapter 1
# THE CHEMISTRY OF LIFE

## Atoms, Molecules, and Chemical Bonding

All matter is built up of simple units called **atoms**. **Elements** are substances that consist of the same kinds of atoms. **Compounds** consist of units called **molecules**, which are intimate associations of different atoms joined in precise arrangements.

Every atom is made up of a positively charged nucleus and a series of orbiting, negatively charged electrons surrounding the nucleus. A simple atom, such as hydrogen, has only one electron circulating around the nucleus, while a more complex atom may have as many as 106 electrons in the various concentric **shells** around the nucleus. Each shell may contain one or more **orbitals** within which electrons may be located. Every atom of an element has the same number of orbiting electrons, which is equal to the number of positively charged **protons** in the nucleus. The balanced number of charges is the **atomic number** of the element. The **atomic weight**, or mass, of the element is the sum of the protons and neutrons in its nucleus. Atoms interact with one another to form molecules, which are held together by chemical bonds resulting from the tendency of atoms to try to fill their outermost shells.

An **ionic bond** is formed when one atom donates electrons to another atom, resulting in oppositely charged ions. **Covalent bonding** occurs when each atom donates an electron to a shared pair. Two or three pairs may be shared forming a double or triple bond.

**Example 1.1** Carbon dioxide ($CO_2$) is a compound in which each of two oxygen (O) atoms forms a double bond with a single carbon (C) atom, which in its  unbonded state has four electrons in its outer shell. In this reaction, two electrons from a C atom join with two electrons from an O atom to form one double bond; the remaining two electrons in the outer shell of the C atom join two electrons from a second O atom to form a second double bond. As a result, the C atom and each O atom has eight electrons in their outer shell, as shown in Fig. 1-1.

$$\overset{\bullet}{\underset{\bullet\bullet}{O}}\mathbin{\vdots} + \overset{\bullet}{\underset{\bullet\bullet}{O}}\mathbin{\vdots} + \times\overset{\times}{\underset{\times}{C}}\times \longrightarrow \underset{\bullet}{\overset{\bullet\bullet}{O}}\mathbin{\vdots\!\times} C \mathbin{\times\!\vdots} \overset{\bullet\bullet}{\underset{\bullet}{O}} \quad \text{or} \quad O{=}C{=}O$$

**Figure 1-1  The formation of $CO_2$.**

In many covalent bonds, the electron pair is held more closely by one of the atoms than by the other imparting a degree of polarity to the molecule. Since oxygen nuclei have a strong attraction for electrons, water ($H_2O$) behaves like a charged molecule, or dipole, with a negative oxygen end and a positive hydrogen end. Many properties of water are based on this polarity.

In the **hydrogen bond** a proton ($H^+$) serves as a link between two molecules or different portions of the same large molecule. H bonds are weaker than covalent bonds, but they play an important role in the 3-D structure of large molecules such as proteins and nucleic acids.

## Osmosis

If we were to divide a container into two compartments using a **semipermeable membrane** and were to place different concentrations of a solution on each side of the membrane, the solute molecules would be unable to pass through the membrane but the solvent molecules would move to the region where they are less crowded. Water would move from less concentrated solute concentrations to more concentrated solute concentrations. This phenomenon is known as **osmosis**.

## Special Case of Water

Water is the single most significant inorganic molecule in all life forms. Because of its polar quality, it promotes the dissociation of many molecules into ions, which play a role in regulating such biological properties as muscle contraction, permeability, and nerve impulse transmission.

Water molecules cohere tightly to one another through hydrogen bonding. This gives water a high surface tension and the property of expanding upon freezing. Water also has a high specific heat and, thus, prevents sharp changes in temperature that would be destructive to the structure of many macromolecules within the cell.

## pH

Acidity and alkalinity are measured by a standard that is based on slight ionization of water. Acidity is determined by the concentration of $H^+$, while alkalinity is a function of the concentration of $OH^-$; therefore, the ionization of water $H_2O$ ( $H^+ + OH^-$) theoretically yields a neutral system. In pure water, dissociation occurs so slightly that at equilibrium, 1 mol (18 g) of water yields $10^{-7}$ mol of $H^+$ and $10^{-7}$ mol of $OH^-$. We may treat the un-ionized mass of water as having a concentration of 1M, since its ionization is so small. Thus

$$\frac{[H^+][OH^-]}{[H_2O]} = \frac{[10^{-7}][10^{-7}]}{1} = K_{eq} = 10^{-14}$$

The meaning of this relationship in practical terms is that $[H^+]$ multiplied by $[OH^-]$ will always be $10^{-14}$, the equilibrium constant for water ($K_{eq}$). Thus, as $[H^+]$ increases, $[OH^-]$ must decrease. The expression **pH** is defined as the negative logarithm (1/logarithm) of the $[H^+]$. Since pH is an exponential function, each unit of pH represents a 10-fold change in $[H^+]$. Neutral solutions have a pH of 7, while the maximum acidity in aqueous solutions is given by a pH of 1 and the maximum alkalinity is given by a pH of 14.

## Don't get confused!!!

The *lower* the pH, the *greater* the hydrogen ion concentration . . . So, a pH of 3 represents $10^{-3}$ mol of $H^+$ ions, but a pH of 2 indicates the presence of $10^{-2}$ mol.

The pH encountered within most organisms and their constituent parts is generally close to neutral. Excess $H^+$ or $OH^-$ produced during metabolic reactions are neutralized by **buffer systems** such as the carbonic acid-bicarbonate ion system of the blood.

## Organic Chemistry

**Organic compounds** are the complex compounds of carbon. The backbone of most organic compounds consists of C chains of varying lengths and shapes to which H, O, and N atoms are usually attached. Each C atom has 4 electrons in its outermost shell and the ability to form double and triple bonds with its neighbors. The organic compounds most associated with basic life processes are carbohydrates, proteins, lipids, and nucleic acids.

**Example 1.2** Among the organic compounds found in nature are the **hydrocarbons**, the molecular associations of C and H, which are non-soluble in water. **Aldehydes** have a double-bonded O attached to a terminal C atom; this carbon-oxygen combination is referred to as a **carboxyl group**. **Ketones** contain a double-bonded O attached to an internal C atom. An **organic alcohol** contains one or more hydroxyl (–OH) groups, and an **organic acid** contains a **carboxyl group** (an –OH and a double-bonded O attached to a terminal C atom).

## Carbohydrates

**Carbohydrates** are hydrates of C with a general **empirical formula** of $C_x(H_2O)_x$. The basic subunit of carbohydrates is a **monosaccharide**, or simple sugar. The most common monosaccharides contain six C atoms and are known as **hexoses**.

**Example 1.3** A typical hexose monosaccharide such as **glucose** consists of a carbon chain to which are attached hydroxyl groups. (Figure 1-2 shows the structural formula for glucose.) These –OH groups confer both sweetness and water solubility upon the molecule.

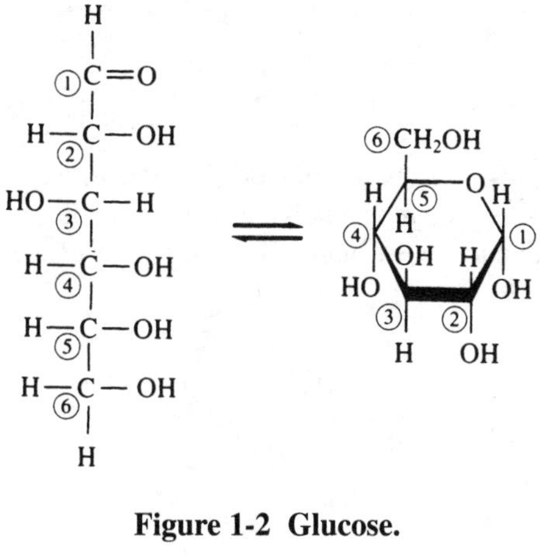

**Figure 1-2 Glucose.**

Monosaccharides may fuse through a process known as **condensation,** or **dehydration synthesis.** In this process two monosaccharides are joined to yield a **disaccharide**, and a molecule of water is liberated (an OH from one monosaccharide and an H from the second are removed to create the COC bond between the two **monomers,** or basic units). Common table sugar is a disaccharide formed by the condensation of glucose and fructose. Condensation may occur many times to yield **polysaccharides.** Breaking the bonds is accomplished by a reversal of the condensation process: water is added back to the molecule. This process is called **hydrolysis.**

## Think about it:

*"carbo-hydrates"* = carbon atoms + water

**Glycogen** is a highly branched polysaccharide chain of glucose units that serves as an energy storage molecule in animals, principally in liver and muscle. In plants, **starch** is the primary storage form of glucose. The primary structural component in plants is **cellulose**, a water insoluble polysaccharide. A structural polymer similar to cellulose, but found in fungi and in the exoskeletons of insects and other arthropods, is **chitin**.

## Proteins

Proteins are a class of organic compounds consisting almost entirely of C, H, O, and N. A protein is a polymer composed of many **amino acid** subunits. The amino acids usually found in proteins show the following structure:

$$R - \overset{\overset{\displaystyle NH_2}{|}}{\underset{\underset{\displaystyle H}{|}}{C}} - C \overset{\displaystyle O}{\underset{\displaystyle OH}{\diagdown}}$$

The COOH (carboxyl) group is characteristic of all organic acids and is attached to the same C as the $NH_2$ group. This C is designated the **carbon** atom. The **R** is a general designation for a variety of side groups that differentiate the 20 different amino acids found in nature. Such properties of a protein as its water solubility or charge are due to the kinds of R groups found in its amino acids.

Amino acids join by expelling a molecule of water. An OH is removed from the carboxyl group of one amino acid, and an –H is removed from the amino group of a second. The resultant bond between the C and N atoms of the carboxyl and amino groups is called a **peptide bond**, and the compound formed is a **dipeptide**. If many amino acids are joined in this condensation process, the result is a **polypeptide**; such chains of amino acids may range from 100 to 1000 amino acids.

# To Summarize

Dehydration = removal of water to synthesize
Hydrolysis = addition of water to break down

It is the same process whether you are building polymers of amino acids or of sugars!

**Primary, Secondary, Tertiary, and Quaternary Structure.** The linear order of amino acids in a protein establishes its **primary structure**. Changes in conformation due to H-bonding between amino acids constitute the **secondary structure** of the protein molecule. Examples are the **α-helix** and the **pleated sheet**. **Tertiary structure** may be influenced by disulfide bridges and charge interactions as well as by hydrogen bonding and results in the 3-D shape of the molecule. Finally, some proteins are composed of two or more separable polypeptide chains. The aggregation of multiple polypeptides to form a single functioning protein is called the **quaternary structure**.

Heat or an appreciable change in pH can lead to alterations in the secondary, tertiary, and quaternary structure of a protein. This is known as **denaturation**. Denatured proteins generally lose their enzymatic activity and may demonstrate dramatic changes in physical properties.

**Remember**

Structure determines function.
This is a common biological theme.

Many proteins are intimately attached to nonprotein organic or inorganic groups to form **conjugated proteins**. Among such nonprotein (prosthetic) groups commonly encountered are carbohydrates (glycoproteins), lipids (lipoproteins), and such specialized compounds as the heme portion of hemoglobin.

**Proteins as Enzymes.** As **enzymes**, proteins serve as catalysts that regulate the rates of the many reactions occurring in the cell. Enzymes, typically characterized by the suffix **ase**, facilitate many biochemical reactions involved in cellular metabolism by lowering the activation energy of these reactions. The catalyzed reaction proceeds at velocities between $10^6$- to $10^8$-fold greater than the uncatalyzed reaction at a given temperature. Enzymes are generally complex globular proteins with a special region in the molecule known as the **active site**. The substance that the enzyme acts upon, the **substrate**, fits into the active site. As a result, the substrate is stretched or distorted and thus more readily undergoes appropriate chemical change. Essentially, the enzyme lowers the resistance of the substrate to alteration and promotes a reaction in which the substrate is changed to its products that allows biological reactions to proceed at relatively low temperatures.

## Lipids

**Lipids** are a class of organic compounds that tend to be insoluble in water or other polar solvents but soluble in organic solvents such as toluene or ether. They consist mainly of C, H, and O. Lipids have much more energy associated with their bonding structure than do the carbohydrates or proteins.

Besides serving as media of energy storage, certain kinds of lipids cushion and protect the internal organs of the body, while others, in the form of a layer of fat just below the skin in many mammals, provide insulation against possible low environmental temperatures.

Among the major classes of lipids functioning within living organisms are

1. the neutral fats (triglycerides),
2. the phospholipids,
3. the steroids, and
4. waxes (found as protective layers on the surfaces of many plants and animals).

## Solved Problems

**Problem 1.1** What is the basis for interaction of atoms with one another?

All the chemical reactions that occur in nature appear to be due to the necessity of atoms to fill their outer electron shells. The noble gases already possess a full complement of electrons in their outer shell and are chemically unreactive.

**Problem 1.2 Diffusion** is the tendency of molecules to disperse throughout a medium or container in which they are found. How does diffusion differ from osmosis; how is it similar?

Diffusion involves movement of *solute* particles in the absence of a semipermeable membrane. Osmosis is a special case of diffusion involving movement of *solvent* molecules (water) through a semipermeable membrane. The two processes are similar in that movement of the molecules in each is driven by their collisions and rebounds with their own kind and proceeds toward areas in which collisions are less likely, namely, areas with fewer molecules of their kind.

# Chapter 2
# CELL STRUCTURE AND FUNCTION

## The Cell Doctrine

All living things are made up of **cells**. The cell is the basic unit of life. Many whole organisms consist of a single cell, while others are a highly complex arrangement of up to trillions of cells. The cell doctrine maintains:

1. All living things are made up of cells.
2. Cells are the units of structure and function.
3. All cells arise from preexisting cells.

# Cellular Organelles

Cells consist of an outer membrane, an interior nucleus, and a large mass of cytoplasm surrounding the nucleus. The functions of the cell are accomplished by specialized structures called **organelles** (see Fig. 2-1).

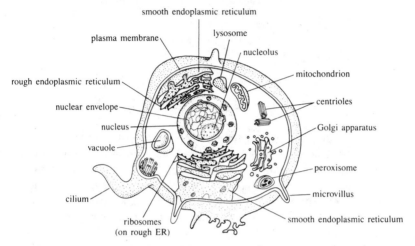

smooth endoplasmic reticulum

plasma membrane

lysosome

nucleolus

rough endoplasmic reticulum

nuclear envelope

nucleus

vacuole

cilium

ribosomes
(on rough ER)

mitochondrion

centrioles

Golgi apparatus

peroxisome

microvillus

smooth endoplasmic reticulum

**Figure 2-1  Structure of a typical animal cell.**

The **cell membrane** or **plasma membrane** is the outer layer of both prokaryotic and eukaryotic cells. It controls the passage of materials into and out of the cell. The **fluid mosaic model** (see Fig. 2-2) proposes a double layer of phospholipids, with their polar ends facing the inner and outer surfaces and the hydrophobic, nonpolar ends apposed at the center of the bilayer. Proteins may reside on the exterior or interior face of the lipid bilayer (**extrinsic proteins**) or may be located in the phospholipid matrix (**intrinsic proteins**); some may be embedded in the bilayer but project through to the exterior, the interior, or both surfaces of the membrane. The plasma membrane is surrounded by a supportive **cell wall** in some organisms. In plants, fungi, and bacteria, this outer coat is composed, respectively, of cellulose, chitin, or a variety of complex carbohydrate and amino acid combinations.

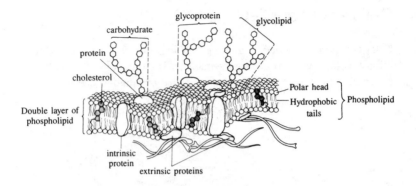

**Figure 2-2 The fluid mosaic model of a membrane.**

The **endoplasmic reticulum** (ER) is a series of continuous membranous channels that traverse the cytoplasm of most eukaryotic cells. The **rough endoplasmic reticulum** (RER) is bound by **ribosomes**, which are associated with active protein synthesis. The **smooth endoplasmic reticulum** (SER) does not contain ribosomes and is associated with the synthesis and transport of lipids or the detoxification of a variety of poisons.

The **Golgi apparatus** is similar in membrane structure to the ER. Its major function is storage, modification, and packing of materials produced for secretory export. Its secretory material is released within secretory vesicles that migrate to the surface of the cell. It may also provide material for the cell membrane.

**Mitochondria** have a double wall: a smooth membrane that forms the outer boundary and an inner membrane that is extensively folded. The mitochondria contain their own DNA and ribosomes and a variety of enzymes, which are involved in the systematic degradation of organic molecules to yield energy for the cell.

**Lysosomes** are full of powerful enzymes used for intracellular digestion. **Peroxisomes** are similar except that the enzymes contained in these organelles are involved in the oxidative deamination of amino acids.

**Flagella** and **cilia** are hair-like projections used for motility. Both derive their motility from **micotubules**, elongated, cylindrical structures assembled from two protein subunits called α- and β-tubulin. The flagella and cilia of eukaryotes contain an outer circle of nine pairs of microtubules that surround a central core of two microtubules. A **basal body**, to which each flagellum or cilium is attached, lies just under the surface of the cell. **Centrioles**, which are also made of microtubules, play a role in the formation of the spindle apparatus, which is an essential feature of both mitosis and meiosis.

**Example 2.1** A human sperm cell has an active beating flagellum at its posterior end. The energy for this activity is derived from a mitochondria-rich midpiece, to which the flagellum is attached. The nuclear payload of the sperm cell is propelled by the flagellum along the female reproductive tract, where it may eventually fertilize an egg.

Cellular support and movement are supplied by the cell's **cytoskeleton**, which comprises three different main types of protein—microtubules, microfilaments, and intermediate filaments. **Microfilaments** are formed from the protein actin. In combination with the protein myosin, they form sliding filaments associated with muscle contraction. The **intermediate filaments** are found, for example, in dense meshworks that give strength to epithelial tissue sheets.

**Vacuoles** are reservoirs within the cell that contain water and salts. Many protozoans have a **contractile vacuole**, which contracts and forces fluid out of the cell. This prevents an accumulation of fluids in organisms that live in fresh water. Vacuoles containing digestive enzymes may also be formed around ingested food particles in a variety of cells. In the cells of many plants, a large **central vacuole** swells against the cell wall, giving the cell a high degree of rigidity, or **turgor**.

Plant cells contain a variety of tiny membrane-enclosed sacs called **plastids** that contain pigments or provide storage space for starch. **Chloroplasts** are included in this group (see Chap. 8).

The **nucleus** is a round or oval body lying near the center of the cell. It is surrounded by a double membrane, the nuclear membrane or envelope. **Pores** provide a route for materials to leave the nucleus. Within the nucleus, one or more nucleoli may be seen. The **nucleoli** are involved in the assembly and synthesis of ribosomes. **Chromosomes** contain the cell's genetic material. The number of chromosomes may differ between species but is constant within a species.

## ✴ Let's Compare...

|  | Animal Cells | Plant Cells |
|---|---|---|
| Cell wall | none | made of cellulose |
| Centrioles | present | not present |
| Vacuoles | less significant | prominent |
| Plastids | not present | in most |

## Tissue Organization

In both plants and animals, groups of similar cells are organized into loose sheets or bundles called **tissues**. Tissues carry out a specific activity and are arranged in discrete structures known as **organs**. Organs carry out a specific function within the organism. A number of organs may be associated as an **organ system**, a complex that carries out some overall function. The sequence of cell → tissue → organ → organ system → organism represents a hierarchical structure in which higher levels integrate the processes derived from lower levels.

## Movement into and out of the Cell

The property of membranes that permits movement of specific molecules across their surface while not allowing movement of other molecules is called **selective permeability**. The internal environment of the cell is carefully maintained by the selective permeability of the cell membrane. If the movement of molecules across the membrane is down the concentration gradient and energy is not used, then transport is **passive**. If the flow is against the concentration gradient, energy must be expended, and the term **active transport** is used. A variety of nonlipid materials pass across the membrane through special **channels** or pores, which facilitate the passage of particular molecules or ions on the basis of their diameter, charge, or ability to form weak bonds between the migrant species and some constituent of the channel. In **facilitated**

**transport** systems, **carriers** combine with small molecules or ions to aid in transport.

Large particles may get into the cell by **endocytosis**. Such particles may attach to membrane receptors. The particle-membrane complex invaginates and pinches off to form vesicles within the cytoplasm. The engulfment of large structures such as bacteria by white blood cells is a type of endocytosis known as **phagocytosis**. **Pinocytosis** is a form of endocytosis in which tiny indentations of the surface permit an interior migration of tiny particles and their surrounding fluid. In **exocytosis**, materials enclosed in a membranous vesicle are brought to and released through the surface membrane.

| Method of Transport | Example |
| --- | --- |
| passive diffusion | water |
| facilitated diffusion | glucose |
| active transport | Na$^+$ |

## Cellular Respiration

**Cell Metabolism.** Cellular combustion of fuels such as carbohydrates, fats, and even proteins releases potential energy stored in the chemical bonds of these molecules. The energy released is stored in ATP molecules, which are broken down by the cell, and the released energy is used for such activities as movement, reproduction, and synthesis. The breakdown of fuel molecules is controlled and occurs in a series of coordinated steps known as a metabolic pathway.

**Anaerobic Pathways.** The first stage in the breakdown of a typical cellular fuel such as glucose involves a metabolic pathway called **glycolysis**. Glycolysis does not require molecular oxygen; therefore, it is an anaerobic process. Glycolysis and **fermentation** refer to essentially the same process, although their end products differ. Although glycolysis involves the breakdown of glucose or related carbohydrates to pyruvic acid, other foodstuffs may "plug in" to the process at a variety of places.

Under anaerobic conditions, pyruvic acid generally reacts with hydrogen to form either ethyl alcohol (most plants and bacteria) or lactic acid (animals and some bacteria). This reduction of pyruvic acid is essential so that $NAD^+$ necessary for glycolysis is regenerated. Without it, glycolysis would stop. However, glycolysis proper ends with the formation of pyruvic acid.

Because each starting glucose yields two trioses, the actual energetic yield of glycolysis is two ATPs.

## Summary of Key Events in Glycolysis

| In | Out |
| --- | --- |
| 1 glucose | 2 pyruvates |
| 2 ATP | 4 ATP |
| 2 NAD | 2 NADH |

**Aerobic Pathways: The Krebs Cycle.** The Krebs cycle is the major aerobic pathway for oxidative degradation of the products of glycolysis. It is also known as the **citric acid cycle** and occurs within the mitochondria. Before entering the Krebs cycle, pyruvic acid is degraded to acetaldelyde (a 2-carbon molecule), with the loss of $CO_2$. The acetaldehyde is then oxidized to acetic acid and attached to coenzyme A (CoA), with $NAD^+$ being reduced to NADH in the process. The acetyl CoA then enters the Krebs cycle.

**Aerobic Pathways: Electron Transport Chain and Oxidative Phosphorylation.** The **electron transport chain** (ETC) represents a series of respiratory pigments of the mitochondrion that function as a "bucket brigade" for the passage of electrons from reduced coenzymes (NADH, $FADH_2$) to oxygen. The passage of electrons down the ETC results in the release of energy which is used to make ATP: two ATP molecules are produced when $FADH_2$ is the electron donor; and three ATP molecules are produced when NADH contributes its electrons.

## Summary of Key Events in Aerobic Metabolism Following Glycolysis Per Glucose Molecule

| In | Out |
|---|---|
| 2 pyruvates | 6 $CO_2$ |
| 4 NAD | 4 NADH |
| 2 FAD | 2 $FADH_2$ |
| | 2 ATP |

**Biosynthesis.** The degradative pathways described above also play a role in the synthesis of vital cellular molecules. For example, the glycolytic process can be reversed (with special enzymatic steps to overcome the irreversible reactions) to form glucose from small molecules. The synthesis of neutral fats requires materials supplied by glycolysis or by phosphorylation of glycerol. The fatty acids of such lipids may be enzymatically chopped into 2-carbon fragments (acetyl CoA) and enter the Krebs cycle for further degradation. Acetyl CoA molecules may also be utilized to synthesize long-chain fatty acids. As can be seen, a vital link exists between catabolism and anabolism within the cell.

Amino acids and proteins may also participate in the metabolic mill. The amino groups can be stripped from their basic carbon skeletons (**deamination**) and attached to different ones. Such metabolic intermediates such as pyruvic acid or α-ketoglutaric acid can be formed from the amino acids alanine or glutamic acid.

## Solved Problems

**Problem 2.1** The endosymbiotic hypothesis states that certain organelles of the eukaryotic cell arose as prokaryotic invaders of eukaryotic precursors. What organelles would you think are included in this hypothesis?

---

## Remember

### Proper Bookkeeping of Aerobic Metabolism

There is a net production of 36 molecules of ATP for each glucose molecule entering glycolysis and proceeding through the Krebs cycle and the electron transport chain. Many texts give 38 molecules of ATP as the total yield, but that does not account for the energy expended when the NADH from glycolysis is brought across the mitochondrial membrane to the ETC. Additional energy is expended to transport ATP from the mitochondria into the cytoplasm.

---

Mitochondria and chloroplasts demonstrate characteristics that are consistent with this point of view. They both contain their own DNA and ribosomes and replicate independently of the rest of the cell. Further, the characteristics of the DNA and ribosomes of these organelles are very similar to those of present day prokaryotes. The existence of many one-celled organisms that contain smaller photosynthetic prokaryotes within their cytoplasm also supports the theory.

**Problem 2.2** Lipids, small molecules, and uncharged particles pass into and out of the cell with relative ease. What characteristics of the cell membrane can be inferred from these observations?

Permeability exists in all membranes and is a function of the cell membrane. The tendency of lipid-soluble molecules to move readily across is due to the large amount of lipid found in the membrane. The greater ease with which smaller molecules enter the cell suggests there are tiny openings in the membrane that confer the property of a sieve on the outer cell boundary. Membranes usually have a charge on their outer surface, and this may explain why uncharged particles may move through more readily than charged particles.

**Problem 2.3** Under aerobic conditions, degradation of the end product of glycolysis (pyruvic acid) produces lactic acid or ethyl alcohol. Why is this step beyond glycolysis necessary?

If glycolysis is to continue in animals, the $NAD^+$ must continually be regenerated. This is accomplished by "dumping" the electons and hydrogen ions of NADH onto pyruvate to form lactic acid and, thus, $NAD^+$ is regenerated. In bacteria, a carbon atom is first removed as a molecule of $CO_2$ from the pyruvate, and the resulting 2-carbon compound is reduced to ethyl alcohol, with a consequent regeneration of $NAD^+$.

# Chapter 3
# THE MOLECULAR BASES OF INHERITANCE

## Introduction

The complex activities of cells are controlled by a set of blueprints (genes) locked in the chromosomes. In essence the blueprints contain information for the synthesis of specific proteins. The nature of the proteins made will determine the structural and functional characteristics of the cell or organism. The properties of **deoxyribonucleic acid** (DNA) can explain the entire agenda of information encoding, processing, replication, and mutability.

# DNA Structure

DNA is a repeating polymer made up of four nucleotides: adenine, guanine, cytosine, and thymine. Using x-ray diffraction photographs of DNA by Rosalind Franklin and the findings of many other investigators, James Watson and Francis Crick developed a double helical model for the structure of DNA in 1953 (see Fig. 3-1).

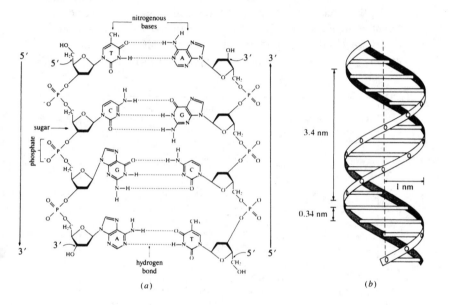

**Figure 3-1  Structure of DNA.**

The hydrogen bonding of the bases (along with hydrophobic interactions between them) holds the double helix together. Adenine is always crosslinked with thymine (by two hydrogen bonds) and cytosine is always linked with guanine (by three hydrogen bonds). Adenine and thymine are therefore said to be **complementary bases**, as are guanine and cytosine. Each sugar forming the backbone of the strand is joined by a phosphate molecule at C5 to the sugar above; a phosphate at C3 joins it to the sugar below. One end of the strand is called the 5' end; the other end is called the 3' end (see Fig. 3-1). Each strand runs opposite in polarity to its complementary strand, thus they are said to be **antiparallel**.

## The Genetic Code

The genetic code for assembling amino acids into proteins (Table 3.1), appears to be universal for all living forms. There must be a minimum of 20 different **codons** to account for the 20 naturally occurring amino acids. If we take three bases as the unit codon, we have $4^3$, or 64, different arrangements possible. The code is **degenerate** in that a number of different codons may represent the same amino acid. Some of the codons may be used as starting or stopping signals during the synthesis of a protein.

| First Position (5′) | Second Position | | | | Third Position (3′) |
|---|---|---|---|---|---|
| | U | C | A | G | |
| U | Phe | Ser | Tyr | Cys | U |
| | Phe | Ser | Tyr | Cys | C |
| | Leu | Ser | (CT) | (CT) | A |
| | Leu | Ser | (CT) | Trp | G |
| C | Leu | Pro | His | Arg | U |
| | Leu | Pro | His | Arg | C |
| | Leu | Pro | Gln | Arg | A |
| | Leu | Pro | Gln | Arg | G |
| A | Ile | Thr | Asn | Ser | U |
| | Ile | Thr | Asn | Ser | C |
| | Ile | Thr | Lys | Arg | A |
| | Met(CI) | Thr | Lys | Arg | G |
| G | Val | Ala | Asp | Glv | U |
| | Val | Ala | Asp | Gly | C |
| | Val | Ala | Glu | Gly | A |
| | Val(CI) | Ala | Glu | Gly | G |

The codons shown are for mRNA, so uracil (U) replaces thymine. C = cytosine; A = adenine; G = guanine. The amino acids encoded for are: Ala = alanine; Arg = arginine; Asn = asparagine; Asp = aspartate; Cys = cysteine; Gln = glutamine; Glu = glutamate; Gly = glycine; His = histidine; Ile = isoleucine; Leu = leucine; Lys = lysine; Met = methionine; Phe = phenylalanine; Pro = proline; Ser = serine; Thr = threonine; Trp = tryptophan; Tyr = tyrosine; Val = valine. CI = chain initiation; CT = chain termination.

**Table 3.1 The genetic code.**

# Protein Synthesis

**Transcription.** The information contained in the DNA sequence is transferred to another molecule called **messenger RNA** (mRNA) by the process of **transcription**.

**Example 3.1 RNA** stands for **ribonucleic acid**. RNA resembles DNA but is single-stranded and has ribose sugars rather than deoxyribose sugars (a hydroxyl instead of a hydrogen is attached at the 2' carbon of the ribose sugar). It is also different from DNA in using the pyrimidine uracil in place of thymine.

Transfer of the information is accomplished by the DNA strand acting as a template for the arrangement of RNA bases with a sequence complementary to that of the DNA. The sequence begins with bases that will later permit the mRNA to attach to a ribosome. The balance of the relatively long mRNA molecule codes for the amino acid sequence of the protein that will be constructed during the translation process and ends with a codon that will signal termination of the synthesis of the protein (CT, see Table 3.1).

Unwinding proteins begin transcription by breaking the H bonds between complementary bases, so that the DNA opens. The enzyme **RNA polymerase** attaches to the **promoter** region of DNA, where synthesis begins. The RNA polymerase catalyzes the formation of mRNA by attaching incoming nucleotides that are complementary to the bases on the exposed DNA template. A termination signal marks the end of the finished mRNA molecule, which is released and soon migrates out of the nucleus. Note that the direction of RNA synthesis is 5'→3'.

The transcription process in eukaryotes is complicated by the existence of long sequences of DNA (**introns**) that do not contain meaningful information for protein synthesis. During transcription, both the introns and the DNA sections that *are* subsequently translated (**exons**) are transcribed into RNA. Before the mRNA leaves the nucleus, the introns are excised, or spliced out, and the remaining sections are annealed yielding functional mRNA.

**Translation.** The next step in the processing of information is the actual production of protein, the **translation** of the code. Translation requires ribosomes, mRNA, and **transfer** RNA (tRNA), which is responsible for carrying amino acids to the mRNA for incorporation into the protein chain. One end of each tRNA binds to a specific amino acid; the other end contains an **anticodon**, a base triplet sequence that is complementary to one of the mRNA codons for the amino acid at the other end. Thus, there are a least 20 tRNAs.

The process of translation begins with the migration of mRNA to the cytoplasm, where the 5' end binds to a ribosome. Essentially, the ribosome "scans" the mRNA in the 5'→3' direction, ensures the tRNAs bring the correct encoded amino acids, and catalyzes the formation of peptide bonds between amino acids. When the ribosome reaches a termination signal, it releases the polypeptide, which may then move through the channels of the endoplasmic reticulum. Several ribosomes may simultaneously translate a single mRNA. Such a configuration of a single mRNA with a number of ribosomes is known as a **polysome**.

The fundamental dogma of molecular biology is summed up as DNA → RNA → protein. This one way flow of information was accepted until the discovery of **reverse transcriptase**, an enzyme associated with certain RNA viruses (**retroviruses**). This enzyme catalyzes the formation of DNA on an RNA template and thus reverses the usual flow of information. Although a special case, the situation in the retroviruses challenged the universality of the fundamental dogma of Watson and Crick.

## Let's practice translation:

| | |
|---|---|
| mRNA | 5'-AUG-GCA-AGG-CCU…UGA-3' |
| polypeptide | met-ala-arg-pro…CT |

Note: AUG acts as a start codon for all polypeptide chains, but it may be cleaved following translation.

# The Concept of Control

Virtually all the DNA in a bacterium like *E. coli* is active gene material that has the capacity to code for as many as 4,000 different polypeptide chains. There are elaborate mechanisms for regulating which genes will be expressed and when. These mechanisms for the regulation of gene expression may be asserted at the level of transcription, translation, or afterward.

Eukaryotic gene regulatory mechanisms are less understood than prokaryotic mechanisms. The number of genes within the total genome of a eukaryote is up to 800 times greater. In addition, for any single eukaryotic cell, as many as 99 percent of all potential genes are shut off. Genes that are actively expressed are found along less deeply staining portions of the chromosome, the **euchromatin**. Deeply stained regions, the **heterochromatin**, generally contain genes that were never or are no longer active.

# DNA Replication

In DNA synthesis, each strand of DNA can serve as a template for a complementary strand. The double helix unwinds and separates, and the individual strands attract their complementary bases (as in mRNA synthesis); each original strand is thus again associated with its complement, and two identical double helices have been created. This is called **semiconservative replication** because each double helix created consists of one "parent" strand and one newly synthesized strand.

# Mutation

Any change in the DNA sequence is termed a **mutation**. One kind of mutation is caused by the addition of an extra base. This **insertion** of an extra base drops the reading frame back one letter (base) from the point of insertion and thus changes all the codons subsequent to it. A similar shift in the reading frame would occur if a base were **deleted** from the gene. In this case the reading frame would be advanced one letter, and all the subsequent codons would be changed. Less likely to eradicate the synthesis of a protein would be a mutation called a **substitution**, in

> # Remember
>
> There are lots of proteins necessary for DNA replication, but they are easy to remember because their names resemble their function.
>
> *Topoisomerases* — cut one of the DNA strands so it can unwind and relieve pressures in the coil
> *Single-stranded DNA binding protein (SSB protein)* — keeps the parent strands separated
> *Helicases* — unwind the double-stranded DNA
> *Primase* — makes primers to initiate chain growth
> *DNA polymerase* — enzyme that attaches nucleotides together in 5'→ 3' direction
> *Ligases* — anneal abutting segments of unjoined DNA
> *Exonucleases* — removes sections of the DNA

which one base is substituted for another. In such an event, one amino acid may be substituted for another because only a single codon is changed. If a new amino acid results (the new codon may encode the same amino acid due to degeneracy) and the new amino acid is similar in its properties to the original one, no resulting damage would be expected. Recently, the phenomenon of **transposition** (jumping genes) has become better understood as another source of genetic variation. In this situation relatively long stretches of DNA jump from one chromosome to another in an unexplained manner.

The agents that cause mutations are known as **mutagens**. Among the most potent of mutagens are various chemicals, ionizing radiation, and ultraviolet light.

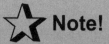

## Note!

Though mutations have a bad rap, they serve as the raw material for evolution. They lead to the differences between individuals that natural selection acts upon.

## Solved Problems

**Problem 3.1** What is a gene?

A **gene** is a unit of information that directs the activity of the cell or organisms during its lifetime. It passes its message along to the progeny when the cell or organism divides or reproduces, so that the gene is also a unit of inheritance. A single gene contains the information necessary to make a single polypeptide chain, which may or may not need to combine with other polypeptide chains to form a functional protein. A gene may also encode a type of RNA other than mRNA.

**Problem 3.2** Discuss the different kinds of RNA present in the cell.

There are three main types of RNA, and each is coded for by specific genes in the chromosomes. All three types are single stranded, but in some cases the strand may fold upon itself to produce a local region of double-strandedness.

One type of RNA is found in the ribosome (rRNA). A second type is mRNA, which is the product of transcription and the template for translation. Transfer RNA (tRNA) represents a third type of RNA that carries amino acids to the ribosome during translation for incorporation into the growing polypeptide chain.

**Problem 3.3** Because DNA replication occurs only in the 5'→3' direction of the new DNA strand and because the two strands of the double helix are antiparallel, uninterrupted replication would seem posible only on one of the template strands. How do you suppose replication occurs from the other strand?

Fork movement (opening of the helix) for the second template strand is in the 5'→3' direction, and therefore the polymerases, which can elongate a growing strand in the 5'→3' direction only and so must always move in a 3'→5' direction on a template, must replicate DNA in a direction opposite to fork movement. Replication cannot be continuous on this strand (the **lagging strand**), since the polymerases must wait for the new stretches on the template to be exposed by the unwinding helix and then work backward from the fork. The result is a series of unconnected segments (**Okazaki fragments**) that eventually are joined to form an uninterrupted strand.

# Chapter 4
# THE CELLULAR BASIS OF INHERITANCE

IN THIS CHAPTER:

✔ *Cell Cycles and Life Cycles*
✔ *Chromosomes*
✔ *Mitosis*
✔ *Meiosis*
✔ *Sexual Reproduction and Genetic Variability*
✔ *Solved Problems*

## Cell Cycles and Life Cycles

**Cell Cycles.** The distribution of nuclear materials into new daughter cells is known as **mitosis**. The apportionment of cytoplasm is called **cytokinesis**. These processes are part of a larger sequence of events known as the **cell cycle**.

The significant vegetative functions of the cells occur during the $G_1$ phase of the cell cycle. These include growth, an increase in the number of organelles, and production of materials for both export and intracellular use. During the S period, the DNA of the nucleus doubles. Following this synthesis stage a second gap, $G_2$, occurs during which materials for the specialized structures required for chromosome movement and cell replication are organized.

**Life Cycles.** In single-celled organisms that reproduce asexually, the cell cycle represents the entire sequence of events occurring during the life of the cell. In organisms that reproduce sexually, the cell cycle is incorporated into a more complex sequence of events. In this sexual process, sex cells (**gametes**) fuse, forming a **zygote**. In most organisms the gametes are known as **sperm** and **egg**. They each have only half the number of chromosomes found in the nonreproductive (**somatic**) cells of the organism. Their union, called **fertilization**, results in a doubling of the chromosome number. This latter condition is called the **diploid** state (designated 2n), while possession of only a single set of chromosomes is known as the **haploid** state (1n). A process of chromosome reduction called **meiosis** intervenes between one sexual event and another.

## Chromosomes

Information for carrying out the activities of every cell is contained within the DNA. In bacteria, a single circular strand of naked DNA makes up the single bacterial chromosome. Some DNA is found as small circlets (**plasmids**). In eukaryotes, the chromosomes exist as numerous linear rodlike bodies. In both prokaryotes and eukaryotes the genes exist as specific stretches of DNA along the chromosome(s). In eukaryotes, the DNA is packaged into coiled **nucleosome** units made of a set of eight basic histone proteins and a few hundred base pairs of DNA. It has also been shown that less than one-tenth of the DNA within the eukaryotic chromosome is actually translated into protein. The prokaryotic chromosome exists as a single unit (is haploid). The eukaryotic chromosomes exist as homologous pairs (are diploid).

Every species has a particular number and characteristic morphology (**karyotype**) for its chromosome sets. Particular genes are found in the same locus of a specific chromosome. Alteration in the number of

## Did you know?

Some strains of viruses utilize RNA as the blueprint material.

chromosomes or in their structure is usually associated with genetic abnormalities.

**Example 4.1** The normal karyotype for humans consists of 23 pairs of chromosomes. The twenty-third pair constitutes the sex chromosomes, which consist of a pair of X chromosomes in the female or an X and Y in the male. In some people a third chromosome 21 is found in the cells; this situation, known as **trisomy**, causes **Down syndrome**.

## Mitosis

**Mitosis** is the process during which the chromosomes are distributed evenly to two new cells that arise from the parent cell undergoing division. During the S phase of interphase, each chromosome will have replicated. The two chromosomal strands (**chromatids**) are identical in their genetic material and are joined at a constricted region called the **centromere**.

Mitosis has four main stages—prophase, metaphase, anaphase, and telophase (see Fig. 4-1). In **prophase**, the nuclear membrane breaks down and the spindle forms. The chromosomes condense and begin to move toward the **equatorial** (middle) plane of the cell. **Metaphase** is characterized by the precise lineup of the chromosomes along the equatorial plane. At the start of **anaphase** the centromeres of each chromosome split so that each chromatid now exists as a separate chromosome. Guided by spindle fibers, the chromatids of each pair are moved to opposite poles. Once the chromosomes reach opposite poles, the last phase of mitosis, **telophase**, begins. During telophase, the cell returns to its premitotic state.

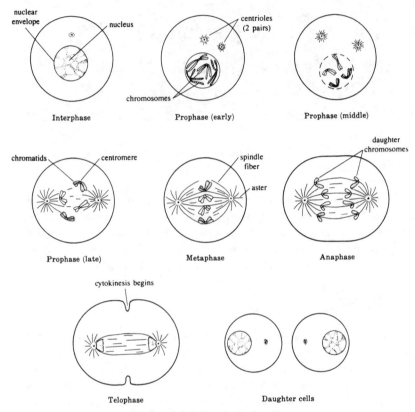

**Figure 4-1 Mitosis (animal).**

With the completion of the nuclear division events, the cytoplasm usually begins its division—a process known as **cytokinesis**. Although accomplished differently in animals and plants, the results are the same: the creation of two separate cells. In animal cells a **cleavage furrow** begins as a puckering along the surface of the cell in the region of the equatorial plane. It extends and deepens until the original cell is completely cut in two. In all higher plants, cytokinesis begins near the middle of the cell with the formation of an internal **cell plate** along the line of the equatorial plane. This plate gradually extends to the surface of the cell and partitions it into two new cells.

# Meiosis

**Meiosis** is the process by which haploid gametes are formed. In animals, the two divisions of meiosis produce the haploid gametes, which eventually unite to form a diploid zygote. In many plant cells meiosis occurs some time after fertilization, with alternating haploid and diploid phases occurring in the plants. More primitive plants spend a greater portion of their life cycle in the haploid (**gametophyte**) stage, whereas more advanced plants are characterized by a dominant diploid (**sporophyte**) stage.

**Meiosis I: Reductional Division.** Meiosis begins in similar fashion to mitosis (see Fig. 4-1): each chromosome replicates in the S phase of interphase, and prophase I begins after $G_2$ with condensation of the doublet chromosomes. The nuclear membrane begins to break down, centrioles move to opposite poles of the cell, and the chromosomes begin to migrate toward the equatorial plane. Spindle fibers start to aggregate from microtubules, and nucleoli disappear. Early in prophase I the homologous chromosomes come together in pairs (**synapsis**). Homologues touch at one or several points; then the chromatids appear to zip together to form an intimate four-stranded structure known as a **tetrad**. In late prophase I, the individual chromosomes from each tetrad start to separate and **chiasmata** may form. Each of the chiasmata formed along the various homologues represents a point at which a section from one chromatid has physically broken off and been exchanged with the corresponding chromatid section on the homologous chromosome. Such an exchange is known as **crossing over**. The metaphase of meiosis I consists of a lining up of *pairs* of homologous chromosomes. During anaphase I whole chromosomes separate with homologues moving to opposite poles. In telophase I, the chromosomes lose their density, a new nuclear membrane forms around each haploid set of doublet chromosomes, and the usual events of telophase ensue.

**Meiosis II: Equational Division.** In the second meiotic division, a haploid set of replicated chromosomes in each new cell migrates to the equator. The centromeres now split, and the sister chromatids of each chromosome migrate to opposite poles. Each of the two cell products of meiosis I will produce two new cells, a total of four haploid cells during the full meiotic process. In some cases, only one functional cell arises from the meiotic process, since in many species each of the two meiotic divisions produces one functional cell and one very tiny **polar body**, which quickly degenerates. The production of gametes in females (**oogenesis**) and males (**spermatogenesis**) is similar in terms of chromosome behavior. However, in the apportionment of cytoplasm to the resultant cells and their modification differences arise.

**Example 4.2** Eggs, or **ova**, are produced from diploid cells called **oogonia**, which are found in follicles within the ovary. These oogonia may enlarge and undergo modification preparatory to meiotic division, at which time they are designated **primary oocytes**. Following meiosis I, a large **secondary oocyte** is produced (containing most of the original cytoplasm) along with a tiny polar body. Both the secondary oocyte and polar body have a haploid complement of chromosomes, but the oocyte contains most of the food material of the cell. During meiosis II, a large cell and a second polar body are produced. This large cell develops into a mature egg, and the polar body disintegrates. Thus, the result of meiosis in female animals is the production of a single, large haploid egg.

In males, diploid **spermatogonia** proliferate mitotically within the testes in the seminiferous tubules. At maturity these spermatogonia undergo modification to become **primary spermatocytes**, the cells that undergo meiosis I to form haploid **secondary spermatocytes**. The secondary spermatocytes each undergo meiosis II to form a total of four haploid cells with single-stranded chromosomes. All four cells are viable and generally have equal amounts of cytoplasm. The cells produced at the end of meiosis are called spermatids. They undergo considerable modification before they are released as functional sperm (**spermatozoa**).

In humans, the female fetus begins creating cells (oogonia) that, upon her birth and development, will ultimately be sequestered as ova in her ovaries. These cells begin meiosis while still in the embryo but are arrested in the prophase I. They remain in "suspended animation" until shortly before fertilization and undergo the second meiotic division only after fertilization. In males, meiosis does not begin until maturity.

## Compare ✔

| Mitosis | Meiosis |
|---|---|
| Prophase shorter | Prophase I longer |
| Single division | Two divisions |
| Occurs in somatic cells | Occurs in germ line cells |
| Products are diploid | Products are haploid |
| Products are genetically identical to parent | Much genetic variation |
| No crossing over | Crossing over |

## Sexual Reproduction and Genetic Variability

One explanation for the ubiquity of sexual reproduction is the variability it provides for the forces of evolution to act on. Variation, in providing flexibility within species, enhances the potential for survival of the species in the face of environmental challenge. Genetic mutation and sexual reproduction greatly increase genetic variability. Meiosis introduces two sources of genetic recombination that add to this variability. The first involves the recombination of whole chromosomes. The second source of variation through chromosomal combination is crossing over.

## Solved Problems

**Problem 4.1** How does cellular reproduction differ in prokaryotes and eukaryotes?

Cell division in prokaryotes is less complex than in eukaryotes, in part because of the presence of only a single chromosome. The complex network of ancillary structures that aid in separating the homologous chromosomes of eukaryotes is unnecessary. As the single, circular strand of DNA replicates, the resulting two circlet strands merely attatch to the plasma membrane. The cell then undergoes cytokinesis, with the membrane and wall growing in, and the two daughter cells separate.

In eukaryotes, the numerous threadlike chromosomes are each replicated and then go through the more complicated process of mitosis. In this process, considerable modification of the chromosomes occurs, and the spindle apparatus and other structures effecting chromosome migration arise.

The special form of cell reproduction that results in reduction of chromosome number (meiosis) is not found in prokaryotes, since they exist in the haploid condition.

**Problem 4.2** Abnormalities in chromosome number give rise to diseases of karyotype. How might these aberrations occur?

Under usual conditions of meiotic division each tetrad separates into constituent homologous chromosomes. If this separation does not occur, all the tetrads may move to one pole, while the opposite pole may receive no chromosomes at all, producing diploid gametes. Should such a diploid gamete unite with a typical haploid gamete, a **triploid** zygote would result. In plants, the formation of triploid and even higher orders of **polyploidy** represents a mechanism for producing new species of the organism in the course of evolution.

More commonly, a single tetrad will fail to separate into its constituent chromosomes. This will eventually result in gametes that have a double dose of one chromosome and others that have no representative for that particular chromosome. The failure of tetrads to disjoin is called **nondisjunction**, and several disorders arise from such phenomenon: Down syndrome (three copies of chromosome #21), Klinefelter syndrome (XXY male genotype), and Turner syndrome (XO female genotype).

# Chapter 5
# THE
# MECHANISM OF
# INHERITANCE

IN THIS CHAPTER:

✔ *Mendel's Laws*
✔ *Linkage and Mapping*
✔ *Sex-Linkage*
✔ *Variations in Gene Expression*
✔ *Solved Problems*

## Mendel's Laws

Gregor Mendel was an Austrian monk born in the early 1800s, whose experiments with garden peas laid a permanent foundation for classical genetics. Mendel showed that heredity involved the interaction of discrete separable factors—a **particulate** theory of inheritance.

Pea plants are extremely useful for genetic studies because they are cheap and easy to procure, produce progeny in a single growing season, and cover their sexual parts with modified petals so that philandering will not intrude upon a careful experiment. Mendel studied seven distinct pea traits, e.g., stem length and seed color.

**Mendel's First Law.** Mendel produced varieties of plants that always self bred true, generation after generation. He crossed strains that were true breeders but of opposite types for each of the seven different characteristics he had selected.

**Example 5.1** Mendel bred a strain of peas that was always tall. He crossed them genetically with a strain that bred true for shortness. In the first filial ($F_1$) generation (progeny of the original, $P_1$, parental cross) he found that all the offspring were tall. Next he crossed all the $F_1$ progeny with themselves. In the resulting $F_2$ generation, Mendel found that there were both tall and short plants. The short trait, which was not expressed in the $F_1$ generation, reappeared in the $F_2$ generation. Of the 1,064 young pea plants in the $F_2$ generation, 787 were tall and 277 were short—close to a 3:1 ratio—with none of the offspring intermediate in size.

Mendel realized that this trait did not run a spectrum of values but existed in two distinct classes, tall or short. He hypothesized that the determinants of a trait also existed as discrete separable factors, a factor for tallness and a factor for shortness. We would now call these factors **genes**. Tall plants had the tall factor, and short plants the short factor.

Mendel also realized that individuals did not have a single factor for any inheritable trait, but a *pair* of factors. One form of the factor (tallness, in this case) tends to be **dominant** over the other (shortness), which is said to be **recessive**; so when both are present, only the dominant one is expressed in the way the offspring looks. This accounts for the fact that in the $F_1$ generation all the offspring were tall; however, lurking in the genetic underpinnings was the factor for shortness.

By keeping track of the alleles in the parents' gametes, it is possible to explain Mendel's results.

**Example 5.2** The gene for height in Mendel's peas exists in two allelic forms. The allele for tall stature will be designated $T$ and that for short stature $t$. The genotype (kinds of alleles present) for the homozygous (same alleles) tall parent would be $TT$; while that for the homozygous short parent, $tt$.

$$P_1 \text{ cross: } TT \times tt$$

The gametes of the tall plants would all have the $T$ allele; the gametes of the short plant, all $t$ alleles.

In the $F_1$ generation the $T$ and $t$ gametes unite to produce individuals with a $Tt$ genotype. Each of these progeny produces gametes, some of which contain the $T$ allele and some the $t$ allele. The various combinations of the $F_2$ can be shown by placing classes of sperm along one axis and classes of egg along its perpendicular. All possible crosses are then shown as the intersections of gametic columns and rows:

|   | $T$ | $t$ |
|---|-----|-----|
| $T$ | $TT$ | $Tt$ |
| $t$ | $Tt$ | $tt$ |

The square formed is called a **Punnett square**. The $F_2$ shows a $TT{:}Tt{:}tt$ **genotypic** ratio of 1:2:1, but a 3:1 **phenotypic** ratio of dominant to recessive pea plants, since the presence of the $T$ allele confers the dominant trait upon the individual.

The gene for height, or any other trait, may exist in two or more alternative forms known as **alleles**, e.g., tallness and shortness. If a pair of alleles in an individual are the same, the individual is said to be **homozygous** for the trait in question. An individual with a pair of contrasting (different) factors is **heterozygous**. The alleles that are present in the genome make up an individual's **genotype**; the individual's appearance is called the **phenotype**.

# Mendel's First Law: Law of Segregation

There exists a pair of alleles for each gene which must separate at gamete formation and then come together randomly at fertilization.

**Mendel's Second Law.** Mendel then studied whether two traits examined simultaneously were inherited independently or were influenced by one another. When he performed crosses involving two traits at a time, he found that the assortment of alleles was completely "democratic." A gamete that received the dominant allele for height during the segregation process might receive the dominant or recessive allele for seed color. All possible combinations of traits could appear in the $F_2$ generation.

**Example 5.3** The $F_1$ progeny are heterozygous for height (tall is dominant) and for seed color (yellow is dominant) : *TtYy*. If independent assortment occurs, each heterozygote can produce four classes of gametes: *TY*, *Ty*, *tY*, and *ty*. A Punnett square can be used to show the possible combinations of these gametes from a dihybrid cross.

|  | *TY* | *Ty* | *tY* | *ty* |
|---|---|---|---|---|
| *TY* | *TTYY* | *TTYy* | *TtYY* | *TtYy* |
| *Ty* | *TTYy* | *TTyy* | *TtYy* | *Ttyy* |
| *tY* | *TtYY* | *TtYy* | *ttYY* | *ttYy* |
| *ty* | *TtYy* | *Ttyy* | *ttYy* | *ttyy* |

There are four possible phenotypes that can result from the independent assortment of the genes for height and seed color: tall plants with yellow seeds, tall plants with green seeds, short plants with yellow seeds, and short plants with green seeds. The Punnett square shows that these are expected to occur in the ratio 9:3:3:1.

# Mendel's Second Law: Law of Independent Assortment
Traits are inherited independently.

**The Chromosome Theory of Inheritance.** Mendel's only paper of his results was published in 1866, but it received little attention from other scientists. In 1900, Mendel was "discovered" by three separate groups, each of whom had independently worked out the law of segregation.

In 1901, William Sutton developed the concept that chromosomes are the physical basis of inheritance. It is the chromosomes that contain a pair of alleles, since each chromosome shares with its colleague an identical locus for the same gene. It is the chromosomes that segregate from one another during gamete formation, and it is the chromosomes of sperm and egg that come together randomly at fertilization.

## Linkage and Mapping

In the early 1900s, Thomas Hunt Morgan found that certain dihybrid combinations produced only two of the four possible classes. Those traits that did not segregate independently were said to be **linked**. Linkage occurs between genes that lie on the same chromosome. Independent assortment of alleles or such genes cannot occur, because during gametogenesis they migrate as a unit.

**Example 5.4** Assume that at the gene locus for flower color a dominant mutation occurs that causes blossoms to be blue, instead of the normal red. If, on the same chromosome, a nearby gene locus that codes for leaf shape has a dominant allele for round leaves, these two alleles, because of their linkage, will always enter gametes together (discounting crossing over), and, whenever blue blossoms are present, only round leaves (not the other, unlinked alleles for leaf shape) will accompany them.

Although only two of the four possible phenotypic combinations would occur for most of the offspring studied, out of hundreds of these offspring a few would show **reassortment** (separation) of the linked traits. Genes that are linked may, during crossing over, break that linkage through an exchange of parts between homologous chromosomes. If alleles $A$ and $B$ are found on one homologue and $a$ and $b$ on the other, linkage dictates that gametes would contain the $AB$ homologue or the $ab$ homologue; a gamete with an $Ab$ or $aB$ combination would supposedly not form.

However, if crossing over occurs anywhere along the length of the chromosome *between* the two gene loci, the exchange of chromatids would interchange one of the loci, while leaving the other locus unmoved. This would produce these new combinations of alleles and permit the unexpected outcomes.

The frequency of recombination of linked genes is an indication of how far apart genes were positioned along the linear chromosomes. The further apart two linked genes are located on their chromosome, the greater the likelihood that crossover events will occur between them.

## Sex-Linkage

In the crosses that Mendel performed, the results did not depend on which parent contributed a particular set of alleles to the zygote. Some traits are **sex-linked**, that is they have genes on the sex chromosomes. Since the male has only one X, a recessive mutation occurring on the X chromosome in the male will be expressed because there is no homologous chromosome to contain an allele that might suppress it. Women with one wild-type allele and one mutant allele are of normal phenotype but are *carriers* for the disease. Half their sons will have the disease and half will be normal.

**Example 5.5** In *Drosophila,* white eye is a mutant recessive trait, while red eye is the wild-type. White-eyed males crossed with homozygous red-eyed females produce offspring all of whom are red-eyed. On the other hand, white-eyed females crossed with red-eyed males produce offspring in which the females are all red-eyed while the males are all white-eyed. Since the gender of the offspring is a factor in the pattern of inheritance, this is a classic case of sex-linkage. All female offspring must receive an X chromosome from the male parent. In the latter cross, since the dominant red-eye allele was on the males' only X chromosome, all females received the dominant allele and were red-eyed. All male offspring must receive an X chromosome from the female parent; in this cross the males received the recessive allele for white eyes and, lacking a homologous allele, expressed the recessive.

---

# Did You Know?

Hemophilia and color-blindness are two common sex-linked characteristics in humans.

---

In 1948, Murray Barr and Dewart Bertram discovered a darkstaining locus in the nuclei of female mammals that was not present in the nuclei of male cells. Dubbed **Barr bodies**, these structures were later found in the cells of men who suffered from Klinefelter syndrome (XXY genotype). Barr bodies represent a highly condensed inactivated X chromosome. Whenever two X chromosomes are present together, only one will exert a genetic effect; the other will remain inactive as a tightly coiled mass of heterochromatin. Inactivation is a random event, so that some of the cells of the female are influenced by the paternal X and others by the maternal X.

## Variations in Gene Expression

Simple dominance is just one of many ways in which genes interact with each other and with their environments. In the case of the four o'clock flower, a cross between purebred red flowers and purebred white flowers produces an $F_1$ with pink flowers. If the $F_1$ generation is crossed, we get offspring which are red, white, and pink in a 1:1:2 ratio. Neither allele (red or white) is completely dominant. This is an example of **incomplete dominance**.

Still another aspect of gene expression is **epistasis**. It involves the effect of the alleles of one gene on the expression of the alleles of an entirely different gene. In **pleiotropy** a single gene exerts an influence on several characteristics. In **codominance** both alleles are fully expressed simultaneously.

## Solved Problems

**Problem 5.1** Yellow haired house mice interbreed and produce progeny with a 2:1 ratio of yellow to nonyellow. When yellow is crossed with nonyellow, a 1:1 ratio of these two classes is obtained. Nonyellows interbreed to produce all nonyellow offspring. How can this be explained?

This would appear to be a case in which yellow is dominant. The anomaly is the yellows seem to be heterozygous but fail to yield the usual 3:1 ratio. The explanation is that the homozygous dominant is a lethal combination causing death before birth. All surviving yellow mice are therefore hybrid. The hybrid cross, which usually gives a 3:1 ratio, is characterized here by a 2:1 ratio since the homozygous dominant does not show up in the final accounting.

**Problem 5.2** In domestic poultry, the character of the comb is controlled by two genes, rose and pea. If the dominant allele *R* is present with a dominant *P*, then a "walnut comb" is produced. If an individual is homozygous recessive (*rrpp*) the comb will be "single." If an *R* is present without a *P*, the comb will be "rose," whereas a *P* without an *R* produces a "pea comb." Determine the phenotypes of the cross *RrPp* X *Rrpp*.

First determine the gametes; then perform the cross.

| Trait | P₁ Cross (Dominant × Recessive) | All f₁ Progeny | f₂ Progeny Dominant | f₂ Progeny Recessive | Total f₂ Progeny | Ratio, Dominant to Recessive |
|---|---|---|---|---|---|---|
| Stem length | Tall × short | Tall | 787 | 277 | 1064 | 2.84:1 |
| Seed form | Round × wrinkled | Round | 5474 | 1850 | 7324 | 2.96:1 |
| Seed color | Yellow × green | Yellow | 6022 | 2001 | 8023 | 3.01:1 |
| Flower position | Axial × terminal | Axial | 651 | 207 | 858 | 3.14:1 |
| Flower color | Purple × white | Purple | 705 | 224 | 929 | 3.15:1 |
| Pod form | Inflated × constricted | Inflated | 882 | 299 | 1181 | 2.95:1 |
| Pod color | Green × yellow | Green | 428 | 152 | 580 | 2.82:1 |

# Chapter 6
# CLASSIFICATION
# OF PROKARYOTES

IN THIS CHAPTER:

✔ *Introduction*
✔ *Archae vs. Eubacteria*
✔ *Bacterial Midgets*
✔ *Bacterial Morphology*
✔ *Origin of Organelles*
✔ *Bacteria and Human Interactions*
✔ *Solved Problems*

## Introduction

The kingdom **Monera** comprises the prokaryotic, single-celled organisms. They possess ribosomes and a naked, circular strand of DNA that serves as a chromosome, but they generally lack membrane-enclosed organelles such as mitochondria, lysosomes, peroxisomes, an endoplasmic reticulum, and a true nucleus. They divide by binary fission. Their fossils have been found in rock strata that are 3.5 billion years old. The monerans include two of the three domains of life: *Archeae* and *Eubacteria*. (The third is *Eukarya*, which is discussed in Chap. 7.)

45

## Archae vs. Eubacteria

**Archae** are probably the oldest of living cells. Their cell walls lack **peptidoglycan**, which is found in the walls of all eubacteria. The lipids in their plasma membrane are branched, differing not only from those of other bacteria but also from those of all other organisms. This unusual lipid makeup is probably related to the extreme environments to which they have adapted. The phototrophic forms use the pigment bacteriorhodopsin instead of bacteriochlorophyll (chlorophyll $a$) used by eubacteria. At the molecular level, their transfer RNA and ribosomal RNA possess unique nucleotide sequences found nowhere else, although structurally more similar to those of eukaryotes than to those of eubacteria!

The archae live in extreme environments where other organisms could not survive. One group, the **methanogens**, lives in bogs and swamplands, where they produce methane through pathways of anaerobic chemosynthesis. These pathways would have been adaptive in the reducing atmosphere of early earth. The **halophilic** (salt-loving) archaebacteria are found in regions of high salinity. A **thermoacidophilic** group has carved out a niche in hot springs and volcanic vents, conditions of high heat and low pH.

The **eubacteria** are extremely wide ranging in their characteristics, and their classification is still quite fluid. Among the recognized major groups are the **purple** and **green bacteria**; each of these is photosynthetic, but they differ from other bacteria in the pathways involved in the process and in the use of $H_2S$ as a source of reducing equivalents rather than water. The **cyanobacteria** carry on photosynthesis in a manner similar to that of the higher plants. A variety of heterotrophic forms are also part of the kingdom. Those bacteria that stain positively with Gram stain are placed in a separate group from those characterized in part by their negative reaction to Gram staining. Some eubacteria contain a flagellum, although it is different in structure from flagella seen in eukaryotic cells.

# Bacterial Midgets

At one time the minute **rickettsiae**, were placed in a separate category. Current opinion among microbiologists is that they are bacteria, although they are not much larger than a virus. They do possess cell structure, and they synthesize their own proteins.

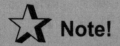 **Note!**

Rocky Mountain fever is a disease caused by a rickettsia carried by ticks.

The **chlamydiae** are another group of minute bacteria. Like the rickettsiae they are capable of reproducing independently, demonstrate organellar organization at least at the ribosome level, and synthesize the coupled enzyme systems that permit the establishment of metabolic pathways.

The smallest of the bacteria are the **mycoplasmas**, a group unique among monerans in that they lack a cell wall.

## Bacterial Morphology

Bacteria are generally one of three shapes—round, rodlike, or spiraled. The **cocci** are perfectly round or eggshaped. Many of the eubacteria have this shape, including the *Pneumococcus*, which causes bacterial pneumonia, and *Streptococcus*, responsible for strep throat. The rodlike bacteria are known as **bacilli**. Bacteria existing as short, helical strands are designated **spirilla**. The causative agent of syphilis, *Treponema pallidum*, is a delicate spiraled moneran.

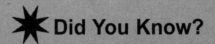

## ✷ Did You Know?

Some forms of chlamydiae cause chronic infections of the human female reproductive tract.

## Origin of Organelles

The **endosymbiotic hypothesis** suggests that chloroplasts, mitochondria, and other organelles were independent bacteria that somehow invaded early cells and set up a permanent housekeeping arrangement within these cells. The invaders became symbionts that profited from the protected environment of the host cell and donated their capabilities to the cell. For example, chloroplasts may have been cyanobacteria whose early lodging within the cell conferred a photosynthetic capacity.

## Bacteria and Human Interactions

Humans depend on bacteria for a variety of necessary functions. Bacteria are the major decomposers of most ecosystems. They break down dead remnants of larger organisms and liberate the constituent molecules and atoms for use by other members of the community. Certain bacteria fix gaseous nitrogen from the atmosphere.

The photosynthetic capacity of most of the members of the cyanobacteria fits them admirably for the role of primary producers in fresh water but especially in the seas. They are probably the major movers in the oxygen revolution that brought aerobiosis to the biosphere some 2.8 billion years ago.

The various types of fermentation carried out by bacteria are particularly useful to humans. Alcohol, acetic acid (vinegar), and acetone are just some of the products produced by bacteria. Bacteria also provide us with cheese and yogurt. The bacterium *Escherichia coli* is a normal inhabitant of our intestines and also a principal tool of molecular biology. Genetic engineering has enabled scientists to insert human genes into bacteria. These bacteria then multiply, cloning the inserted gene, to provide large amounts of such vital proteins as insulin, interferon, and growth hormone.

Although most bacteria enhance the quality of life in the ecosystem and within human communities, they also present a negative aspect as the causal agent of many diseases. These diseases range from leprosy and tuberculosis to typhoid fever and bacterial pneumonia. At the same time, bacteria have been sources of many antibiotics used in the fight against bacterial infections.

## Solved Problem

**Problem 6.1** Spores are usually found among the bacilli. A spore arises within a single bacteria cell as an **endospore**, which is enclosed by a highly durable, impenetrable spore coat. The endospore has enough material from the larger cell to ensure its survival and capability to develop into a new bacterial cell. What function would you say spores serve?

The endospore is hardier than the parent cell and so, represents a strategy for survival of the cell under adverse conditions. Spores can lie dormant for many years and are highly resistant to heat and cold, chemicals and dessication. Spore formation raises the reproductive potential of a bacterial species. In terms of the challenge to human populations, spore formation enhances infectivity, virulence, and parasitic persistence of bacteria. Even treatments such as steam exposure or strong disinfectant may fail to eradicate spores.

**Problem 6.2** Describe several lines of evidence supporting the endosymbiotic theory.

One line of evidence derives from the prokaryotic nature of many cellular organelles. Basal bodies, mitochondria, and chloroplasts contain their own genetic blueprint, are roughly the size of bacteria, and contain their own ribosomes. Another line of evidence derives from present instances of bacteria that have invaded eukaryotic cells and remained there permanently. Similarities between the chlorophylls of the chloroplast and of cyanobacteria support the idea that cyanobacteria are progenitors of the chloroplast. Furthermore, the interior mitochondrial membrane is similar in structure and function to the bacterial plasma membrane.

# Chapter 7
# CLASSIFICATION OF EUKARYOTES

IN THIS CHAPTER:

✔ *Introduction*
✔ *The Kingdom Protista*
✔ *The Kingdom Fungi*
✔ *The Kingdom Plantae*
✔ *The Kingdom Animalia*
✔ *Solved Problems*

## Introduction

The eukaryotes are the third domain of life. Eukaryotes are organisms that are made up of cells that contain a nucleus and other membrane-bound organelles. Four of five kingdoms in traditional classification system are eukaryotic: *Protista*, *Fungi*, *Plantae*, and *Animalia*.

## The Kingdom Protista

The kingdom *Protista* includes all the eukaryotic unicellular species. Some of these organisms are animal-like (**protozoans**), others resemble

51

plants (**algal protists**), and still others demonstrate the characteristics of fungi. Divisions within the kingdom are based on functional characteristics. As with the monerans, taxonomy is in a state of flux, and different classification schemes are found in a variety of biology texts. The protists evolved about 1.6 billion years ago. They are extremely complex; their cells show even more diversity than is found among the cells of the multicellular kingdoms. It is believed that they gave rise to the fungi, higher plants, and multicellular animals.

**Protozoans**. The **protozoans** are heterotrophic organisms found in every major habitat. Some are freeliving, whereas others exist as parasites within the bodies of animals. Protozoa as a division has been divided into five major phyla (Table 7.1).

| PHYLUM | CHARACTERISTICS |
|---|---|
| Mastigophora (zooflagellates) | Flagella, no cell wall, parasitic (*Trypanosoma*) |
| Sarcodina (amoebas) | No cell wall, but some shells (*Foraminifera*) |
| Sporozoa | Parasitic, no means of locomotion, complex life cycles depending on hosts (*Plasmodium*) |
| Ciliata | Parasitic, no means of locomotion, complex life cycles depending on hosts (*Paramecium*) |
| Opalinda | Enteric parasites, ciliated, two or more nuclei |

**Table 7.1 Protozoan phyla.**

**Algal Protists**. Plantlike protists (**algae**) are subdivided into six phyla (Table 7.2). Virtually all the members of this half-billion-year-old group are photosynthetic and occupy fresh or saltwater habitats.

| PHYLUM | CHARACTERISTICS |
|---|---|
| Euglenophyta (Euglena) | Unicellular, many heterotrophic, a few autotrophic, complex structures, flagella, eyespot |
| Pyrrophyta (dinoflagellates) | Unicellular, pair of flagella, thick cell walls, complex type of meiosis |
| Chrysophyta (mostly diatoms) | Chlorophylls $a$ and $c$, produce a yellow-brown carotenoid (fucoxanthin), store food as fats, oils and laminarin, walls of silica, diatoms encased in double shell |
| Chlorophyta (green algae) | Extremely diverse, chlorophylls $a$ and $b$, mostly freshwater, complex sexual and asexual stages, unicellular and multicellular forms |
| Phaeophyta (brown algae) | Multicellular, seaweed, chlorophylls $a$ and $c$, fucoxanthin, store calories as oils and laminarin, alternation of generations |
| Rhodophyta (red algae) | Seaweeds, only chlorophyll $a$, red due to phycoerythrin, multicellular, reproduce sexually, alternation of generations |

**Table 7.2 Algal phyla.**

**Fungi-like Protists.** Protists in this group consist of two heterotrophic groups of slime molds. The *Myxomycota* are plasmodial slime molds. They are highly pigmented amoeboid cells that alternate between a "multicellular" aggregate and individual cells. The aggregate stage consists of a large, **coenocytic** (having many nuclei) mass and is called a **plasmodium**. The *Acrasiomycota* are cellular slime molds. Their aggregation phase is multicellular rather than coenocytic. When food is in short supply, the individual cells aggregate but individual membranes persist and each cell can be distinguished.

The *Oomycota*, which include water molds and some rusts and mildews, are superficially similar to fungi, but the diploid stage is dominant in their life cycle, their cell walls are made up of cellulose, and they possess flagella.

# The Kingdom Fungi

**Fungi** are eukaryotic, heterotrophic, and, except for yeasts, multicellular (or multinucleate). The fungi are at least 400 million years old. They obtain their food by absorption rather than by ingestion. They secrete their digestive enzymes outside their bodies and then absorb the products of digestion. Most fungi possess cell walls made of **chitin**, an amino-containing polysaccharide. They all lack flagella and are restricted in terms of motility. **Yeasts** are unicellular fungi. Molds and mushrooms are other examples of fungi.

**Fungal Structure.** Fungi consist of a tangled mass of multibranched threads called **hyphae** (see Fig. 7-1). These hyphal filaments are incompletely subdivided by **septa**. In most fungi the septa are porous and permit cytoplasmic flow from one "cell" to another. Other groups have coenocytic structures. The entire filamentous mass is called a **mycelium**.

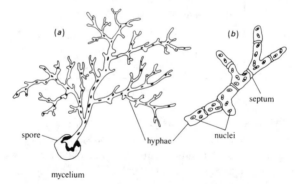

**Figure 7-1 The major features of fungal structure:
(a) mycelium; (b) septate hyphae with multinucleate cells.**

**Divisions of Fungi**. The fungi fall into four basic divisions:

1. The *Zygomycota* (conjugation fungi) occupy a terrestrial habitat, invading the soil or decaying organic matter. The division is named for the tough resistant **zygospores** that are formed when haploid gametes fuse, usually under adverse conditions. The diploid

zygospore then undergoes meiosis, and one or more of the meiotic products give rise to a new haploid mycelium.

2. The *Ascomycota* (sac fungi) include yeasts, some mildews, the ergot, and *Penicillium*. The name of the group comes from the presence of a reproductive sac called an **ascus**, which is formed during the sexual cycle of all its members.

3. The *Basidiomycota* (club fungi) include the mushrooms and toadstools and a variety of puffballs. Here, too, the hyphae are compartmentalized by septa. In most representatives an extensive underground hyphal mass sporadically sends up vertical hyphal fruiting bodies in which the spores are formed. The reproductive structures found in the fruiting bodies are club-shaped.

4. The *Deuteromycota* comprises all the forms in which a sexual cycle has not been discovered.

The lichens are often separated from other fungi. These intimate mutualistic associations of an algal or cyanobacterial autotroph with a fungus are of great ecological significance.

**Reproduction**. Most fungi are haploid through much of their life cycle. Haploid spores form in the **sporangia** of the **sporangiophores** (tips of specialized hyphae). Since these hyphae are haploid, no meiotic process is necessary to produce the spores that form within and burst from the sporangia. Asexual reproduction in yeasts involves budding, in which a new, smaller cell pinches off from a parent cell.

Sexual reproduction usually occurs when food supplies are low or optimal conditions of moisture and temperature do not exist. Fungi from different divisions (except *Deuteromycota*) have unique specialized structures for sexual reproduction.

## Friends or Foes

While we would rather do without ringworm, athlete's foot and vaginal yeast infections, we depend on fungi as effective decomposers and for the production of antibiotics, many foods and alcoholic beverages.

# The Kingdom Plantae

Plants are multicellular, autotrophic organisms that have successfully invaded terrestrial environments. They probably arose from the algal division of *Chlorophyta*. Opportunities for active photosynthesis were considerably enhanced when plants established themselves on land. As plants migrated to drier regions, the reproductive strategies were also modified (see Chap. 8).

More than 400 million years ago, the ancestral forms of today's land plants began to spread through the terrestrial environment and evolve adaptations that facilitated their survival. A major division into two separate lineages occurred early in the colonization of the land. One group was the **bryophytes**, and the other, far more numerous in the modern era, the **tracheophytes** (vascular plants).

## Bryophytes

Bryophytes consist of three extant groups: **mosses, liverworts**, and **hornworts**. Two-thirds of all bryophytes are mosses. Although the bryophytes have developed some protective structures, they are not eminently suited to a terrestrial existence. They require moist environments for their reproductive cycles. Although they do have processes that resemble the roots of higher plants, and even demonstrate chlorophyll-laden scales, they are clearly only a stage in the evolution of higher plants.

The mosses are both more numerous and more visible than other bryophytes. Lacking the internal support structures of higher plants, mosses tend to spread extensively but grow close to the ground. Like other bryophytes, they show a dominant gametophyte (haploid) generation and a dependent sporophyte (diploid) generation. Most mosses are dioecious but some are monoecious.

The liverworts are named for their flat, lobular appearance, resembling the lobes of the liver.

The hornworts are a minor group of bryophytes, possessing some features of higher plants. They get their name from a hornlike sporophyte that grows up from a flattened gametophyte.

# Vascular Plants

Vascular plants differ from the bryophytes in their greater adaptiveness to land environments. The body of vascular plants is divided into separate specialized parts or organs (see Chap. 8): roots, stems, leaves, and cones or flowers.

Possible transition forms in the evolution of vascular plants from multicellular algae belong to the division *Psilophyta*. These plants possess only some of the features of vascular plants, but a primitive vascular system can be discerned.

**Club Mosses and Horsetails.** The club mosses, or lycopods, belong to the division *Lycophyta*. They share a peculiar characteristic with the *Psilophyta* – the gametophyte generation is nonphotosynthetic and must rely on symbiotic fungi for nourishment. The lycopods have both true roots and leaves. The sporophyte generation is dominant. Spores that develop into the gametophyte are formed on special leaves called **sporophylls**.

Another division of seedless plants is *Sphenophyta* (horsetails). They are mostly small, herbaceous (nonwoody) plants. The stems are usually hollow and jointed. Leaves typically form in whorls at each joint. Sporangia are carried in groups at the ends of a central stem.

**Ferns.** The ferns (*Pterophyta*) are the most extensive and numerous of the seedless plants. The leaves are both broader and more vascularized than in the bryophytes or sphenophytes.

# Seed Plants

The development of the seed represents the height of adaptiveness to a terrestrial environment in the plant kingdom. In all seed plants the sporophyte is dominant and the gametophyte is reduced to a dependent structure within the archegonium. Further, the flagellated sperm of the lower plants is supplanted by a process of pollination, achieving independence from water as a vehicle of fertilization. The seed houses the embryo in a tough coat, further relieving the dependence on water.

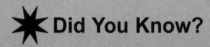

**Did You Know?**

The development of the seed was a major terrestrial adaptation! Why?

— protection from severe temperatures and drought
— easy dispersal to new locales
— growing embryo packaged with food supply

**Gymnosperms**. The four divisions of **gymnosperms** are quite distinct. The *Cycadophyta* resemble palms and grow in warm climates. The *Ginkgophyta* consists of a single species, the **ginkgo**. It is dioecious, resistant to pollution, and well-known for the foul odor emitted by the female in the spring. The *Gnetophyta* include three groups of tropical or desert plants: *Gnetum* (a vine), *Epedra*, and *Welwitschia* (both shrubs). The *Coniferophyta* (conifers) are the most conspicuous of the gymnosperms, especially in colder climates. The most prominent feature is the reproductive structure known as the **cone**.

**Angiosperms**. The flowering plants (**angiosperms**) are the most successful and widespread of all plants. The reproductive structures of the angiosperms are flowers that completely enclose the future seeds. Species that are pollinated by insects have large, magnificently colored petals and are often pleasantly scented.

## The Kingdom Animalia

Animals are multicellular, eukaryotic organisms that are characterized by their nutritional habits — they eat (ingest) other living organisms. Animals that prey on other animals are known as **carnivores**. Others ingest plants and are classified as **herbivores**.

Humans belong to a subphylum of the phylum *Chordata* known as *Vertebrata*. The vertebrates, or backboned animals, only make up 5 percent of the animal kingdom, but they figure prominently in the lives of humans. All other animals are classified as invertebrates.

## Subkingdom Parazoa: The Sponges

The animal kingdom is usually subdivided into groupings designed to reflect the evolutionary relationships of major lineages. The sponges are generally placed in a separate subkingdom known as *Parazoa*, while all other animals, supposedly derived from a separate protistan modification, form the subkingdom *Eumetazoa*.

The sponges are sessile (permanently attached, hence nonmotile), sacklike organisms belonging to the phylum *Porifera*. Pores cover the surface of the sponge body. Water is sucked in through these pores, moves through the body's interior cavity (**spongocoel**), and exits through the **osculum**. Food particles in the water are filtered out by special cells, known as **collar cells**, or **choanocytes**.

Although sponges are multicellular animals, they demonstrate less integration and specialization of function than other animal groups. They lack tissue organization, and their cells are the primary units of structure and function. The body is two layered: an outer epidermis and an inner sheet consisting mainly of choanocytes. A gelatinous compartment, the **mesohyl**, separates these layers. Spiny **spicules** may be deposited in the mesohyl, giving the sponge support.

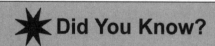

# ✳ Did You Know?

Most sponges are marine, but a few species inhabit fresh water. They vary greatly in size and shape and color.

## Radiata: Cnidaria and Ctenophora

Above the level of the sponges, animals can be broadly divided into two classifications on the basis of body symmetry: radial and bilateral. The radially symmetrical phyla, the *Radiata*, possess a central mouth around which the rest of the organs are radially arranged.

The *Radiata* consist of two phyla: *Cnidaria* and *Ctenophora*. Each of these phyla consist of individuals with two layers of cells, radial symmetry, and an inner gastrovascular cavity. The gastrovascular cavity serves a digestive function. Food is taken in through the mouth, and what is undigested passes out again through the mouth. The cnidarians and ctenophores can ingest large food particles and pass the digested nutrients throughout the body. The *Radiata* display a tissue level of organization. Specialized nerve cells make up a **nerve net**, which may permeate the entire body of the organism. Epitheliomuscular cells make up a contractile tissue.

**Cnidaria**. A striking feature of the cnidarians are the **cnidocytes**. When appropriately stimulated, the cnidocyte "pops" up to sting and lasso prospective prey. The cnidarians are grouped into three major classes and one minor class: *Hydrozoa* (hydras), *Scyphozoa* (jellyfish), and *Anthozoa* (corals and sea anemones) are the major classes, the *Cubozoa* (sea wasps) the minor class. In most hydrozoan life cycles distinct polyp and medusa stages alternate. In the scyphozoans, the medusa stage clearly predominates and the **mesoglea** (matrix between endoderm and ectoderm) is thick. The anthozoans ("flower animals") are the largest and most varied of the cnidarians; they completely lack a medusa stage and consist of large, fleshy cylindrical bodies. They are responsible for the coral reefs found in warm ocean waters.

**Ctenophora**. The ctenophores are similar to the cnidarians in terms of their radial symmetry and two-layered body plan; however, they lack cnidocytes and do not demonstrate any alternation of body form (polyp to medusa) during their life cycle. The mesoglea may contain muscle cells that show independent properties of contractility. The phylum gets its name from the eight rows of fused ciliary plates that run longitudinally along the body. The cilia in these plates (combs) beat synchronously to propel the animal in the water. Usually these comb jellies (another common name) are passive passengers of ocean tides and stronger currents, many are luminescent, and all capture their food with the aid of sticky cells lining the tentacles.

## Bilateria: The Deuterostomes and Protostomes

All the remaining members of the animal kingdom probably arose from a common ancestral form that demonstrated bilateral symmetry. The *Bilateria* have a **head** (anterior end) and **tail** (posterior end). The head shows an increasing complexity of sensory structures in the more recently evolved groups. All the bilateral forms possess a third primary germ layer known as **mesoderm**. All the *Eumetazoa* above the level of *Radiata* are differentiated in terms of the presence or absence of a true **coelem**, a body cavity completely lined with mesoderm.

The coelemate *Bilateria* soon diverged into two major branches in the animal kingdom. The **deuterostomes** form the anus from the blastopore where the **protostomes** form the mouth from the blastopore. Deuterostomes demonstrate radial, indeterminate cleavage, whereas protostomes tend toward spiral, determinate cleavage.

| Deuterostomes | Protostomes |
|---|---|
| Echinodermata | Mollusca |
| Chordata | Annelida |
| | Arthropoda |

The coelem endows an organism with several benefits. The gut is separated from the rest of the body, thus peristaltic movements do not interfere with locomotor contractions of muscles in the body wall. The fluid of the coelem may serve as a hydrostatic skeleton. Also, the membranes of the coelem provide sheets of support for the internal organs.

## Acoelomates: Platyhelminthes

Acoelomates lack a true coelom. This group includes the flatworms, belonging to the phylum *Platyhelminthes*. The **platyhelminths** are simple, but they exemplify an organ level of complexity. Four classes include:

1. *Turbellaria* are free-living flatworms that feed on small organisms. Their digestive cavity is branched and extends throughout the body. An example is *Planaria.*
2. *Monogenea* are parasitic, have a simple life cycle, and are hermaphroditic.
3. *Trematoda* are parasitic flukes that have many intermediate hosts.
4. *Cestoda* are parasitic tapeworms consisting of a head (**scolex**) and repeating segments known as **proglottids**, each of which contains male and female reproductive structures

## Pseudocoelomates: Rotifera and Nematoda

*Rotifera* and *Nematoda* possess a **pseudocoelom**, in other words, their body cavity is not completely lined with mesoderm. **Rotifers** are tiny and, despite their multicellularity, may be smaller than some large amoebas. They contain a characteristic ridge of beating cilia (**corona**) that surrounds their mouth. **Nematodes** (roundworms) are free-living; however, such diseases as hookworm and trichinosis arise from parasitic roundworms.

## Protostome Coelomates

**Annelida.** The phylum *Annelida* comprises worms whose bodies are divided into segments. This segmentation is clear externally but also exists internally in the form of membranes (septa) that subdivide the interior of the worm. Annelids possess a closed circulatory system with enclosed vessels (arteries, capillaries, and veins). A common example is the earthworm.

**Mollusca.** *Mollusca* ("soft-bodied") are tremendously diverse in terms of form and life style. Most are shelled, but some have lost the calcareous shell. All classes except the *Bivalvia* (clams and oysters) possess an extensible, raspy tongue called a **radula**. Rows of strong teeth along this usually flattened structure serve as a hacksaw when the radula is extended and pulled back over a food-laden area. Most of the mollusks have an open circulatory system in which large spaces stand between arteries and veins. Examples include snails and squids.

**Arthropoda**. In terms of numbers, diversity of species, and distribution in the biosphere, the *Arthropoda* are the most successful of all animal phyla. The five classes of arthropods ("jointed feet") comprise well over a million species, and many new species continue to be discovered. Arthropods are covered by a protective **exoskeleton** that contains chitin. The body is usually segmented, with the segments giving rise to various types of appendages. Most arthropods have powerful compound eyes, and the coelom of arthropods is considerably reduced. Arthropods contain an open circulatory system. There are five major classes of arthropods:

1. *Arachnida* have four pairs of legs. They respire by means of tracheae or **book lungs**. Some poison their prey. The spiders spin webs to ensnare unwary victims.
2. *Crustacea* are primarily marine and utilize gills for respiration. All have two pairs of antennae on their heads serving as sensory probes and in defense or food acquisition.
3. *Diplopoda* are herbivores and are almost entirely terrestrial. Each body segment has two pairs of legs.
4. *Chilopoda* are all carnivores and are particularly adept at rapid running.
5. *Insecta* has more species than all other animal organisms combined. They are the most successful invaders of the terrestrial environment. They probably owe their adaptiveness to their small size, wings, compound eyes, metamorphosis, and a social system. The body of most insects is divided into three parts: a head, thorax, and abdomen. The sexes are separate, and fertilization occurs during copulation, the insertion of the male penis into the vagina of the female. Insects are valuable plant pollinators, but they also destroy crops and spread disease.

To help remember the characteristics of each class, consider common examples:

Arachnids — spiders and scorpions
Crustaceans — crabs and barnacles
Diplopods — millipedes
Chilopods – centipedes
Insects – wasps and fleas

## Deuterostome Coelomates

**Echinodermata**. The adults of the phylum *Echinodermata* (e.g., sea stars) demonstrate radial symmetry, but are bilaterally symmetrical in early development. They are marine and share a common basic structure. A **central disk** is circled by five radiating arms. A system of **tube feet** extends from **radial canals** in each arm. A **ring canal** in the disk is the hub of this closed watervascular system. This system functions in locomotion, excretion, and respiration.

**Chordata**. The nonvertebrate phylum *Chordata* consists of two subphyla: the *Urochordata* (tunicates), which inhabit the ocean bottom, and the *Cephalochordata* (e.g., lancelet). They share, with the vertebrates, the three basic characteristics of all chordates:
1. a notochord, a flexible but sturdy rod of fibrous tissue just ventral to the neural tube
2. a hollow dorsal nerve cord derived from ectoderm
3. pharyngeal gill slits
    Features of the subphylum *Vertebrata* include a skeletal backbone and cranium, which enclose the spinal cord and brain; a high degree of **cephalization** (the specialization of the anterior end of the nervous system into a complex brain with associated specialized sense organs); and segmentation of the muscles of the trunk into somites during development.
    Of the seven classes of vertebrates, three are fishes. The remaining four are termed **tetrapods** because they possess two pairs of limbs.
1. The *Agnatha* prey on other fish by attaching to their outer surface and boring through to the internal organs with their jagged tongue.
2. The *Chondrichthyes*, like the agnathans, have skeletons made of cartilage.
3. The *Osteichthyes* are the bony fish, which are found in both fresh and salt water.
4. The *Amphibia* are only partially successful as terrestrial inhabitants. Many amphibians are dependent on water for sexual activity. In frogs, the tadpole undergoes metamorphosis to become a lung-bearing, adult land-dweller. Many amphibians, however, respire principally through a moist surface membrane rather than through their lungs.

5. The *Reptilia* all have scales on a rough, waterproof skin that prevents drying out. Other modifications for successful terrestrial life include functional lungs; an enclosed (cleidoic) egg in which the embryo can absorb yolk and develop within a protective bath of amniotic fluid, protected by a shell; internal fertilization; and a variety of behavioral patterns that enable the reptile to survive extremes of temperature and shortages of food.

6. The *Aves* (birds) have been aptly described as "flying lizards." In addition to the fact that they probably descended from dinosaurs, they possess several features in common with reptile.

7. The *Mammalia* share with insects an extreme mastery of the terrestrial habitat. Mammals have hair (fur), a four-chambered heart, a diapragm, are endotherms, and have the ability to produce milk for their young. Most mammals possess very complex brains and show a highly sophisticated behavior pattern, including play. There are three subclasses of mammals: *Protheria* (lay eggs, lack teeth, have beak), *Metatheria* (have pouch and live young), and *Eutheria* (have placenta and a prolonged pregnancy).

**Consider representative examples of each class:**

*Agnatha* – hagfish and lampreys
*Chondrichthyes* – sharks and rays
*Osteichthyes* – trout and bass
*Amphibia* – frogs and salamanders
*Reptilia* – alligators and lizards
*Aves* – sparrows and chickens
*Mammalia* – dogs and chimpanzees

## Solved Problems

**Problem 7.1** Criticize the statement that the cells of *Protista* are simpler than those of the higher plants or animals.

Actually, protistan cells are far more complex than their counterparts among the multicelled plants and multicelled animals. In a single cell, *Protista* carry out all the functions of an entire organism. Cells in plants and animals need not be involved in all the functional properties of the entire organism, since cells can specialize.

The most complex of all cells are found among the protozoans. Specialized structures exist for ingestion and digestion, motility, water balance, and reproduction. The ciliates probably demonstrate the most complex organelles, some of which are unique to that group.

The many variations of cell division found among the *Protista* illustrate the complexity of protistan calls. Mitosis usually occurs, but it varies greatly in a way that is not paralleled in any other kingdom.

**Problem 7.2** Many fungi resemble protists or even plants. What is the justification for placing fungi in a separate kingdom?

At one time the fungal forms that are currently placed in a separate kingdom were grouped as a division of the plant kingdom. However, fundamental differences distinguish the fungi from both plants and protists. All fungi are heterotrophs, which obtain their food by absorption from outside. Although many fungi store some cellulose in their cell walls, the predominant cell wall constituent is chitin, clearly distinguishing fungi from plants. Unlike many plants and some colonial protists, the fungi do not have flagellated zoospores, nor are the vegetative cells of the mycelium motile.

**Problem 7.3** How do angiosperms differ from gymnosperms?

First, the seeds of the angiosperms, a more recently evolved and highly successful division of the tracheophytes, are enclosed within a protective chamber, the ovary. A ripened ovary containing seeds is called a fruit.

The seed also forms differently in angiosperms. One sperm nucleus from the pollen tube unites with an egg nucleus to produce a zygote. A second sperm nucleus unites with two haploid nuclei in the gametophyte (embryo sac) to produce the triploid endosperm, an important food source within the seed for such seeds as corn.

In gymnosperms, pollination can only be windborne. In flowering plants, pollen may be transferred by wind or by animals.

A major internal modification of the angiosperms is the developments of specialized xylem cells, the **vessels** and **fibers**. The vessels are particularly significant because they are large-bore columnar cells that join end to end. When their inner cellular contents degenerate, they collectively form long tubes that greatly facilitate the passage of water in the plant. Fibers, on the other hand, function solely to provide support. In conifers the single xylem elements, the **tracheids**, represent a more primitive condition.

**Problem 7.4** The arthropods (especially insects) are regarded as the most successful of all animal groups. Why do you suppose this is?

(1) They have the greatest number of species and individuals, thus are more successful at survival than other groups. (2) Their great diversity signals successful evolutionary development; the exoskeleton protects against both dessication and predators, while still allowing free movement. (3) They have coevolved with plants and other animals to ensure survival by mutualistic interactions.

# Chapter 8
# PLANT STRUCTURE AND FUNCTION

## Plant Structure

**Roots.** Water and dissolved nutrients move into a plant through the **roots**. Thin extensions of the cells at the surface of the root, **root hairs**, actually absorb the water and bring it to the conducting tissue within the root. Materials move into the root by diffusion, osmosis, and even active transport.

The root is formed by a series of concentric cylinders. The outermost ring is a single-celled layer of **epidermis**. Next is a thick ring of parenchyma called the **cortex**. The round, loosely packed cells of the cortex have thin walls, facilitating the passage of water and minerals through the cells. The innermost layer of the cortex is a single ring of tightly packed cells, the **endoderm**. Some of the cells of the endoderm have noticeably thinner walls than others; these are called **passage cells** and are involved in the transport of water and minerals to the inner core of the root, the **stele**. The stele is bound by a ring of cells called the **pericycle**. These cells are meristematic; that is, they give rise to secondary cells. In this case, they form secondary **xylem** and **phloem**, the two tissues involved, respectively, with transport of water and nutrients. The xylem and phloem are the primary components of the fibrovascular bundle, or stele. Their arrangement within the stele varies considerably between dicots and monocots (see Fig. 8-1).

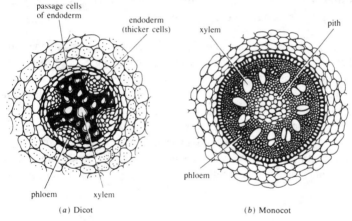

**Figure 8-1  Fibrovascular bundle of root.**

**Stems.** The **stem** connects roots to leaves. It often comprises most of the plant and may be involved in supporting the position of the leaves, carrying out photosynthesis, transporting raw materials and finished primary and secondary photosynthetic products, and storing food materials.

The stem and its branches, along with the leaf system, make up the **shoot**. If the stem remains relatively short with extensive branching throughout its length, it is classified as a **shrub**. Taller perennial plants with thick trunks showing little branching at the base are called **trees**. In woody plants an outer layer of cork-laden cells (**bark**) forms as a watertight shield around the stem. Openings (pores) of the bark called **lenticels** afford exchanges of gases between the internal stem cells and the atmosphere.

Leaves are attached to the stem in a characteristic fashion. The point of attachment is called a **node**. Usually, a leaf is attached to the stem through a thin stalk called a **petiole**. Stems grow through specialized growing structures called **buds**. Buds may give rise to branches of the stem, or they may be specialized to produce flowers.

Stems may broadly be divided into woody and herbaceous. Woody stems are characteristic of trees and are usually found in dicots. Herbaceous stems remain soft and frequently carry on photosynthesis. They are characteristic of most monocots and many dicots.

Many different types of stems have evolved. **Rhizomes** are stems that grow laterally underground.

**Example 8.1** The white potato tuber is a thickened underground stem that is an important repository of stored food.

**Leaves.** Leaves are generally thin, flattened structures that amplify the photosynthetic capacity of the plant. In cross section (see Fig. 8-2), the leaf consists of an upper and lower epidermis. Beneath the upper epidermis is a layer of **palisade mesophyll**, followed by a layer of **spongy mesophyll**. Both mesophyll layers consist of parenchymal cells that are thin-walled, rich in chloroplasts and capable of highly intense photosynthetic activity (discussed later in this chapter). The lower epidermis is generally spotted with a rich array of **stomata**, slitlike openings to the outside. Each stoma is surrounded by a pair of epidermal **guard** cells. A waxy **cuticle** generally covers the leaf.

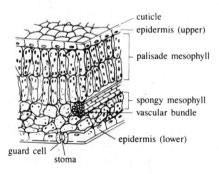

**Figure 8-2 Leaf cross section.**

In **dicots** a main vein, usually in the center of the leaf, breaks up into a complex crisscrossing pattern known as **netted venation**. In the **monocots** a regular pattern of parallel venation, in which veins of similar bore run longitudinally through the leaf in parallel fashion, is found.

**Xylem.** Water and dissolved minerals and gases enter the plant through the root hairs. Since the osmotic pressure within these hairs is generally greater than in the surrounding soil, the inflow of fluid creates a push in the root region, which is called **root pressure**. This contributes to the continuous flow of fluid along the xylem throughout the roots and stems of the plant as does **transpirational pull** from the leaves, the negative pressure, or suction, created by the evaporation of water from leaf surfaces. **Adhesion**, or the attraction of water molecules to the walls of their container, is believed to assist in pulling the water up the stem.

**Phloem.** The phloem transports food materials synthesized in the leaves to all parts of the plant. Since leaves are most abundant in regions distant from the trunk or stem, phloem flow is generally toward the stem and roots. A variety of substances move along the protoplasm of the phloem, but sucrose is usually the most abundant.

---

### *What* is being transported *where*?

Xylem → water up
Phloem → food down

---

## Photosynthesis

**The Process. Photosynthesis** is the process by which high energy organic molecules are produced from simpler components by green plants and other autotrophic organisms in the presence of light energy. In the process of photosynthesis, **photons** (unit packets) of light are captured by specific pigment molecules, such as **chlorophyll**. Electrons within these pigment molecules are excited by the absorbed photons, and these excited electrons eventually liberate their energy to the cell as they fall back to the unexcited ground state. This energy is used to reduce carbon dioxide to a carbohydrate.

**Example 8.2** The photosynthetic reaction is essentially a reversal of cellular respiration. In respiration, energy is released when molecules such as glucose ($C_6H_{12}O_6$) are oxidized to $CO_2$ and $H_2O$. The released energy is stored as ATP. In photosynthesis, energy from the sun is absorbed by pigment systems within the chloroplast, and this energy is utilized first to produce ATP and then to form a sugar molecule. In the process, oxygen gas is liberated.

# You Need to Know

Photosynthesis may be summarized as follows:

*Light Reaction* – light required, NADPH and ATP produced, occurs in thylakoid membrane

*Dark Reaction* – NAPDH and ATP used, $CO_2$ reduced to a carbohydrate, $O_2$ liberated, occurs in stroma

**The Location: The Chloroplast.** The outer and inner membranes of the **chloroplast** (Fig. 8-3) are congruent in shape and lie close to each other. The fluid-filled space enclosed by the inner membrane is called the **stroma**. Distributed throughout the stroma is the **thylakoid membrane system**. This is a network of channels that forms **grana**, stacks of flattened sacks. The components of the light reaction are generally tightly bound to this thylakoid membrane and are accessible to molecules found in the stroma, where the reactions of the Calvin-Benson cycle (dark reaction) occur.

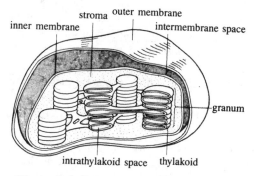

stroma  outer membrane

inner membrane  intermembrane space

granum

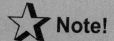

intrathylakoid space    thylakoid

**Figure 8-3  Structure of a chloroplast.**

A chemiosmotic system is responsible for the generation of ATP during the course of the light reaction. Protons are pumped into the inner thylakoid compartment, creating a proton gradient between the inner thylakoid space and the stroma. $H^+$ ions migrate out of the thylakoid space through an ATP synthetase system attached to the thylakoid membrane.

## ⭐ Note!

The $C_4$ pathway, or **Hatch-Slack pathway**, is a modified metabolic pathway believed to be a mechanism for survival by plants in hot, dry environments.

## The Nitrogen Cycle

Plants, like all life forms, require nitrogen. Although 79% of the atmosphere consists of elemental nitrogen, plants cannot incorporate it directly. Intermediate processes must first convert the nitrogen to usable forms.

Nitrogen-fixing bacteria are able to combine atmospheric nitrogen with hydrogen to form ammonium ions ($NH_4^+$). The ammonia is then released into the environment, where it can be taken up and used by plants. Alternatively, nitrogen-fixing bacteria may live mutualistically within special nodules in plants such as clover, alfalfa, and legumes (beans and peas). These bacteria fix nitrogen and supply the resulting ammonia directly to the plant. Ammonia absorbed by plants is converted to amino acids and other nitrogen compounds through the mediation of the enzyme nitrogenase.

Ammonia created by nitrogen fixation may undergo further processing, called **nitrification**. In the first step, bacteria convert the ammonia to nitrites ($NO_2^-$), which then are converted to nitrates ($NO_3^-$) by other bacteria.

**Nitrogen fixation** introduces new nitrogen to the cycle; however, decomposition acts on organic sources of nitrogen, that is, nitrogen that has already been incorporated into living organisms.

All these processes—the introduction of elemental nitrogen through nitrogen fixation and nitrification; the restoration of $N_2$ to the atmosphere through denitrification; the recycling of organic nitrogen through decay, ammonification, and nitrification—are all elements of the **nitrogen cycle**.

## Interactions of Plants with Their Environment

Plants detect changes in their environment and react to these changes in specific ways. Responses may involve movement (but usually not locomotion), alteration of growth patterns, developmental phenomena, or a change in the state of particular plant structures.

**Tropisms.** A **tropism** is an invariable growth response to an environmental stimulus occurring in plants and primitive invertebrates. Tropisms are named for the eliciting stimuli and described as positive if growth is toward the eliciting stimulus and negative if growth is directed away from the stimulus. Responses to gravitation (**geotropism**), water (**hydrotropism**), light (**phototropism**), pressure, touch (**thigmotropism**), etc., are crucial to the survival of a plant. A number of tropisms may be demonstrated simultaneously.

**Example 8.3** The roots of a plant grow toward the origin of gravitational force (center of the earth). Therefore, they are positively geotropic. On the other hand, the shoot grows away from the gravitational source. It is characterized as negatively geotropic.

**Photoperiodism. Photoperiodism** refers to the ability of plants to respond to relative periods of light and darkness. Specific responses, such as flowering, are elicited by long periods of light in some cases and by short periods of light in others.

# Plant Hormones

A **hormone** is a chemical mediator produced in one part of a living organism that has metabolic effects throughout the organism as it travels through the vascular system.

**Auxins** play a significant role in a broad variety of plant behaviors and growth patterns. They are involved in (1) the suppression of lateral buds along the stem, (2) the development of root and shoot systems, (3) the growth of the fruit, (4) the dropping off of leaves and fruit (**abscission**), (5) the division of cells in the cambium, and (6) the development of new structures such as adventitious roots.

**Cytokinins** are a class of hormones that stimulate cell division in plants. They interact with auxins to determine the differentiation of meristematic tissues. They are necessary for the formation of such organelles as the chloroplast and may play a role in flowering, developing fruit, and breaking dormancy in seeds.

The **gibberellins** increase plant height and alter both vigor and nutritional yield.

A unique hormone, because it exists as a gas rather than a liquid, is **ethylene**. It is usually associated with relatively rapid ripening of fruit.

# Plant Reproduction

In plants, two life forms, a haploid gametophyte and a diploid sporophyte generation, alternate with one another through time. In algae, the gametophyte form is usually prominent. In bryophytes, the sporophyte is usually attached to and dependent on the gametophyte. Bryophyte sperm (encased within the **antheridia**) are motile and require moisture to swim to the eggs, which are located within the **archegonia**. The zygote develops in the archegonial receptacle.

In ferns, the dominant phase is the sporophyte generation. The gametophyte exists as a **prothallus**. It is **monoecious** (both sexes on one plant). Sperm are released and must swim to the egg cells within the archegonia. The fertilized egg develops into a sporophyte that is parasitic upon the tissues of the archegonia at first. Then a true root develops and extends beyond the parent plant. On the underside of most fern leaves, round clusters known as **sori** form. Sori contain **sporangia** (spore cases). The cycle is complete when each spore germinates into a new haploid prothallus.

The life cycle of flowering plants (**angiosperms**) is particularly suited to withstand the challenges of a terrestrial life style. Only the sporophyte generation is visible. The haploid gametophyte exists as a parasite within the tissues that flower. The significant feature of the group is the flower, a complex reproductive organ. The innermost whorl, or layer, comprises the female organ, the **pistil**, which is made up of a sticky **stigma** resting on a long **style** leading to the **ovary** and **ovules**. **Stamens**, male organs, surround the pistil. External to the stamens are the **petals**. The outermost layer is the **calyx**, which is composed of the tough **sepals**, which completely enclose the bud of the developing flower.

A mature **embryo sac**, the female gametophyte of a flowering plant, arises within the ovule and contains an egg cell and an **endosperm nucleus**. A pollen grain (containing a divided nucleus), which develops in the **anther sac** of the stamen, is carried to the stigma, and elongates into a **pollen tube** (the male gametophyte). It grows down the pistil stalk. Once the tube has reached the embryo sac, **double fertilization** occurs. One sperm nucleus joins the egg to produce a zygote, which develops into a seed, and the other sperm nucleus unites with the endosperm nucleus, which grows into the **endosperm**, or seed food. Although an endosperm develops in **gymnosperms** (non-flowering seed plants), double fertilization is only found in angiosperms. The seed develops within the ovary, or ripened fruit, of the flower.

## Solved Problems

**Problem 8.1** The stomata are tiny lengthwise openings concentrated on the lower surface of most leaves. They are enclosed by a pair of guard cells, which resemble two bow-legged sausages. Just above the stoma and within the leaf is large air space. Unlike other cells of the lower leaf epidermis, the guard cells contain chloroplasts. What mechanism does this suggest for opening and closing of the stomata?

The guard cells function through osmotic changes in their cytoplasm. During the daylight hours, when the conditions are optimal for active photosynthesis, the guard cells pile up sugar molecules. These sugar molecules increase the osmotic pressure, which leads to an uptake of water and an increase in the turgor (internal pressure) of the cell. When the cells start to swell, they pull apart so that the stomatal opening (pore) increases in size. This permits $CO_2$ and water vapor to pass into the leaf and oxygen, a by-product of photosynthesis, to pass out. Thus, the leaf opens its interior to the atmosphere only when gas exchanges necessary for photosynthesis occur.

When conditions for photosynthesis are poor, the guard cells do not produce sugars and osmotic pressure is reduced. With the loss of turgor pressure, the guard cells become limp, the walls partially collapse, and the opening is occluded by the now flaccid cell walls. This prevents water loss at a time when photosynthesis is not occurring at appreciable levels.

**Problem 8.2** If the critical photo period for a short-day plant is 13 hours of daylight, will it flower when there are 15 hours of daylight?

A short-day plant requires a *minimum* level of *darkness* in order to flower. It follows then that, under natural conditions, the plant may also be said to have a *maximum* level of *daylight*, beyond which it will not flower, since there will not be enough hours of darkness remaining in the 24 hour period. In this example, the critical level of daylight is 13 hours, and any level of daylight beyond this will not allow the plant sufficient hours of darkness. The plant will therefore not flower if there are 15 hours of daylight.

# Chapter 9
# INTERCELLULAR COMMUNICATION

IN THIS CHAPTER:

✔ *Introduction*
✔ *Hormones*
✔ *Endocrine Systems*
✔ *Nervous Systems*
✔ *The Neural Impulse*
✔ *The Synapse*
✔ *Brain and Spinal Cord*
✔ *Autonomic Nervous System*
✔ *Solved Problems*

## Introduction

Two types of integration exist within animal organisms. The **nervous system** is a communication network that affords rapid and specific connections within the body. The nerve impulse is the means by which electrical messages travel from one region to another. The organizing and directing influences of the nervous system reside in the brain and spinal cord of the central nervous system.

78

The nervous system's type of integration, in which impulses travel along well-defined tracts, is in sharp contrast to the second type of integration, the **endocrine system**. **Endocrine glands** are localized secretory structures that make and release **hormones** directly into the general circulation. These hormones are then broadcast to all parts of the body and may produce far-reaching effects on a variety of structures.

In some cases, a combination of neural and hormonal mechanisms functions in intimate associations to achieve integration. Such **neurohumoral** mechanisms are particularly prominent in the functional axis that joins the brain and the pituitary gland.

## Hormones

**Hormones.** Hormones are usually regarded as chemical secretions that are produced in one part of an organism by a specialized endocrine structure and that have profound metabolic effects on **target** structures at distant sites in the body. Hormones may actually inhibit certain processes while stimulating others. Hormones may exert their effects by (1) altering gene function, (2) directly affecting metabolic pathways, or (3) controlling the development of specific organs or their secretory products.

**Hormone Action.** Hormones exert their effects on target tissues by an alteration of the metabolic activity of specific cells or by interacting with the genome to alter gene activity. Hormones must either penetrate the cell or set into motion a chemical chain of events from an attached position on the membrane.

Many hormones attach to specific receptors on the cell membranes of target cells and invoke the aid of a "**second messenger**" in the cytoplasm. The second messenger often activates enzymes that are part of a system whose final step produces the actual end action of the hormone. Generally, this mechanism for hormone action is faster than the mechanisms involving modulation of gene action.

### Remember
Calcium ions and cyclic AMP are common second messengers.

Nonpolar hormones, such as steroids, cross the plasma membrane and form a complex with a receptor in the cytoplasm. This complex moves into the nucleus, and binds directly to the chromatin of specific genes, where a stimulatory or inhibitory effect upon the genetic machinery occurs.

## Endocrine Systems

**Invertebrate Endocrine Systems**. The best known endocrine systems in invertebrates are those associated with growth and metamorphosis in insect groups. Each successive molt (from **larva** to **pupa** to **adult**) is induced by a hormone called **ecdysone**, secreted by the **prothoracic glands** located behind the head. Ecdysone production is stimulated by brain hormone, a peptide produced in **neurosecretory cells** of the forebrain. Although ecdysone induces each molt, it is the blood levels of **juvenile hormone**, produced by the **corpora allata** in the hindbrain, that determine what the stage following the molt will be.

In a variety of invertebrates, hormones control sexual cycles and often are directly involved in the shedding of eggs. All the arthropods demonstrate rather extensive endocrine systems, which play a role in water balance, migration of pigments involved in protective coloration, and growth.

**Vertebrate Endocrine Systems**. The role of endocrine glands in the maintenance of homeostasis in vertebrates has been studied extensively. A sampling of several major glands and their corresponding hormones is listed in Table 9.1. Their secretions are of four basic types: (1) proteins, (2) less complex peptides, (3) catecholamines, and (4) steroids. Figure 9-1 shows the location of the major endocrine glands in the human body.

| SOURCE AND HORMONE | PRINCIPAL EFFECT |
|---|---|
| Pyloric Mucosa of Stomach — Gastrin | Secretion of gastric juice |
| Pancreas — Insulin Glucagon | Glycogen formation and storage Conversion of glycogen to glucose |
| Kidney — Angiotensin | Increases blood pressure |
| Testes — Testosterone | Secondary male characteristics |
| Ovaries — Estrogen Progesterone | Secondary female characteristics Pregnancy |
| Thyroid — thyroxine | Oxidative metabolism |
| Adrenal Medulla — Epinephrine | "Fight or flight" response |
| Hypothalamus — Releasing factors ADH | Regulates hormone secretion Water reabsorption |
| Anterior Pituitary — Growth Hormone Follicle Stimulating Hormone Luteinizing Hormone Prolactin | Growth Growth of follicles and seminiferous tubules Conversion of follicles to corpus lutea, sex hormone secretion Milk secretion |
| Posterior Pituitary — Oxytocin | Contraction of uterine muscles, milk release |

**Table 9.1 Vertebrate endocrine glands and hormones.**

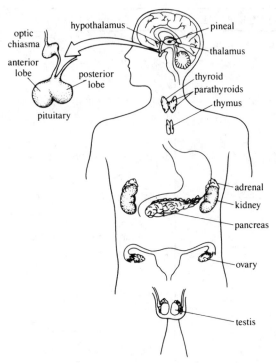

**Figure 9-1 Major endocrine glands in humans.**

# Nervous Systems

The nervous system is based on the capacity of cells and whole organisms to respond to internal or external changes in the environment called **stimuli**. The specific reaction elicited by a stimulus is termed a **response**. These reactions are usually very rapid.

In most nervous systems interconnecting fibers form a communications network through which signals in the form of an electric current flow. The generation of such a **nerve impulse** is called **excitation** and is the result of a localized flow of ions.

Stimuli produce impulses in nerve cells (**neurons**) whose endings are exquisitely sensitive to these stimuli. Such impulses, which move toward the central axis of the nervous system, are called **sensory**, or

**afferent**, impulses. Those impulses that move from the central axis to excite responses by glands or muscles are termed **motor** or **efferent** impulses.

**Receptors and Effectors**. **Receptors** respond to specific kinds of stimuli. Sometimes a receptor is an elaborate sense organ, such as the retina of the eye.

**Effectors** are the end structures that bring about response. They are usually muscles whose contraction produces the response, or they may be glands that secrete particular substances as a result of stimulation.

Sensory organs have special receptors which allow an organism to be aware of its environment. These include receptors specific for light, sound, gravity and movement, taste and smell, touch and pressure, and heat and cold.

**Reflexes** are simple pathways in which information is directly transmitted from the afferent neuron to the effector. The reflexes are mechanisms for maintaining appropriate posture, regulating blood pressure, and orienting the body to environmental conditions that threaten the organism.

**Phylogenetic Comparison of Nervous Systems**. In early multicellular organisms, such as sponges, there is little in the way of a coordinating system. In the *Cnidaria*, communication is accomplished with a **nerve net**. In some cnidarians an additional **nerve ring** exists, which permits a greater complexity of neuromuscular responses. The comb jellies (*Ctenophora*) also have a nerve net, but in this phylum evidence of localized differentiation exists. Among such hints of higher levels of specialization are an **oral ring** around the mouth and a series of eight neural strands immediately below the combs.

In flatworms (*Platyhelminthes*) an organ level of organization is apparent. At the anterior end are two lobes of concentrated nerve tissue that make up the brain. Extending posteriorly are two **nerve cords**. These structures can process information coming from sensory cells at the surface and permit more complex behaviors than those of more primitive organisms.

In mollusks further complexity is achieved by increased **cephalization** (concentration of nerve cells at the head end) and a greater number of knots of nerve cell bodies known as **ganglia**, which are scattered through the nervous system. An array of diverse sensory structures also characterizes this phylum.

The earthworm (*Annelida*) also demonstrates a complex nervous system. A variety of sensory cells are concentrated at the head end, but sensory organs are not prominent. Two large ganglia make up the brain. A double, but fused, nerve cord runs along the body on the ventral surface. In each segment of the earthworm a ganglion buds off the nerve cord to coordinate sensory and motor impulses.

A remarkable degree of cephalization is found in arthropods, particularly the insects. Most prominent of the sense organs in the anterior region are simple or, in some groups, compound eyes. A "ladder" type of nervous system lies along the ventral surface—a double nerve cord punctuated by ganglia lying along its length. Coordination of the delicate movements of the appendages is a function of the ganglia of each segment affording a high degree of decentralization of motor function.

Although the echinoderms are relatively advanced by many criteria, many do not have clearly defined nervous systems.

**The Neuron**. The functional unit of the nervous system in both invertebrates and vertebrates is the **neuron** (see Fig. 9-2). The **dendrites** are projections that carry impulses *toward* the **cell body**, a thicker region of the neuron containing the nucleus and most of the cytoplasm. The **axon** is a long projection that carries impulses *away* from the cell body. Many axons and even dendrites may combine to form a single **nerve**.

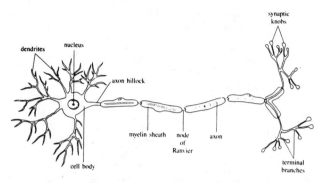

**Figure 9-2  A model of a typical motor neuron.**

The neurons are supported, in both a mechanical and a metabolic sense, by **glial cells**. In the outlying neurons of the **peripheral nervous system**, which carry impulses to and from the central nervous system, the supporting tissue consists of **Schwann cells**, which wrap around the axon in a many-layered insulating cover called the **myelin sheath**. This fatty, membranous sheath affords a rapid and highly efficient transmission of impulses.

## The Neural Impulse

When a neuron is not conducting an impulse, it is said to be in the **resting state** (see Fig. 9-3). A **resting potential**, a difference in charge, exists between the inside and the outside of the membrane. A higher concentration of sodium ions exists outside the membrane, while there is a higher concentration of potassium ions inside.

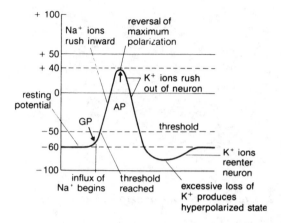

**Figure 9-3 Electrical changes during a nerve impulse.**

These concentration gradients are maintained by the impermeability of the resting membrane to $Na^+$ and the action of a $Na^+/K^+$ pump that, driven by ATP, transfers $Na^+$ to the outside and pumps $K^+$ in. A potential difference inside relative to outside of roughly 60 millivolts (mV) exists across the membrane. The natural tendency to correct this energetically unstable imbalance is the driving force behind the nerve impulse.

When a neuron is stimulated, the point of stimulation suddenly becomes permeable to sodium ions, which rush in, **depolarizing** the membrane (that is, eliminating the potential difference), as the incoming positive ions balance the negative internal charge. Enough $Na^+$ rushes in to make the inside of the membrane temporarily positive (see Fig. 9-3).

This local shift of charge constitutes the neural impulse, or **action potential**. It triggers depolarization of the adjacent area. This process continues as a wave of depolarization down the length of the axon. The impulse is thus not actually transported anywhere but like a wave of water is *recreated* at each point.

When the action potential reaches a maximum (about +40 mV), the membrane suddenly again becomes impermeable to $Na^+$. At the same time, $K^+$ is pumped out, until it balances the number of sodium ions that rushed in and the membrane is **repolarized**. This efflux of positive ions restores the resting potential. There is a brief overshoot of negativity. The $Na^+/K^+$ pumps restore the original gradients. Until the membrane reaches its resting potential again, it is incapable of developing a new action potential; the membrane is said to be in its **refractory period**.

Each neuron has a threshold of stimulation below which it will not fire. Above this threshold all stimuli evoke a depolarization of common intensity. This is known as the **all-or-none principle**. Instead of evoking stronger action potentials, stimuli of increasing strength elicit multiple firings, with the frequency increasing with the strength of stimulation.

## The Synapse

The point at which an axon and a dendrite associate is called a **synapse**. In invertebrates possessing a nerve net the axons and dendrites touch each other at such junctions, so that passage of a nerve impulse across the synapse is an electric event. As a result, the nerve net fires as a unit.

In the typical mammalian synapse (see Fig. 9-4), a definite gap (the **synaptic cleft**) exists at the junction. Movement of the nerve impulse across this gap is primarily a chemical event mediated by neurotransmitters such as acetylcholine, $\gamma$-aminobutyric acid, norepinephrine, and serotonin.

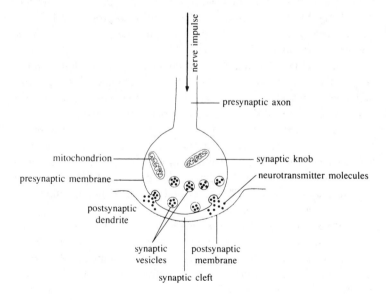

**Figure 9-4  A typical mammalian neural synapse.**

The axon ends in many small **synaptic knobs,** containing a number of **synaptic vesicles,** which are filled with neurotransmitters. When an action potential reaches the knob, voltage-sensitive calcium gates in the knob's presynaptic membrane open, permitting an influx of $Ca^{2+}$. This results in the spilling of neurotransmitters into the synaptic cleft where they bind to special receptors in the postsynaptic membrane of the apposing dendrite or cell body.

To prevent continued random firing of a neuron, enzymes degrade the neurotransmitter remaining in the cleft. Simple synapses also occur between neurons and muscle fibers.

# Brain and Spinal Cord

The brain is encased in a bony skull and is further protected by three specialized membranes or **meninges.** The **cerebrospinal fluid** (CSF) is contained within these and provides both cushioning and a supply of ions for the brain and spinal column.

The vertebrate brain is divided into three basic regions: the hindbrain, midbrain, and forebrain. The **hindbrain** is chiefly concerned with involuntary, mechanical processes. The **midbrain** processes the visual and auditory information and is involved in behavioral patterns in the lower vertebrates.

The **forebrain** includes the hypothalamus, which along with its hormonal responsibilities controls such crucial parameters as heart rate, blood pressure, and body temperature, as well as many fundamental drives such as hunger, thirst, sex, and anger. The **thalamus** provides connections between many parts of the brain and may also control moods and feelings. The **reticular formation** controls the level of arousal, or wakefulness. The largest and most complex part of the forebrain in humans is the **cerebrum.** The cerebrum receives and analyzes stimuli from sensory organs, controls speech and analytical thinking, body perception, memory, and learning.

The **spinal cord** extends from the brain stem to the lower back and is almost completely enclosed by the vertebrae of the spinal column. It consists of ascending tracts conducting sensory impulses toward the brain and descending tracts carrying impulses downward to motor neurons within the spinal cord.

> # Remember
>
> Central Nervous System =
> brain + spinal cord
>
> Peripheral Nervous System =
> everything else!

## Autonomic Nervous System

The **autonomic nervous system** comprises two subsystems. The **sympathetic nervous system** prepares the body for emergency situations, the responses associated with "fight or flight." The heartbeat is accelerated, blood pressure is increased, the blood sugar level rises, and blood moves from vessels within the trunk to those of the arms and legs to support fighting or running away. On the other hand, the **parasympathetic nervous system** maintains those conserving functions that restore the organism during periods of tranquility: eating, sexual activity, urination.

## Solved Problems

**Problem 9.1** How do hormones differ from enzymes?

Enzymes are almost always protein. Hormones may be proteins, shorter peptides, single amino acids, derivatives, or steroids. Enzymes may be synthesized within a cell to function there or may be secreted to the exterior through active transport and even pass along a duct to a specific locale. Hormones are released directly into the bloodstream and are distributed throughout the body where they may exert their effect on target tissues. Enzymes are generally highly specific, catalyzing a single reaction. Hormones may demonstrate different but significant actions on different tissues in the same organism. Hormones play a key role in the maintenance of their levels through negative feedback involving tropic hormones and the releasing factors of the hypothalamus; this is not paralleled by enzymes

**Problem 9.2** What mechanism causes the action potential to be an all-or-none response?

The size of the action potential is determined by the amount of sodium that pours into the depolarizing neuron. The amount of sodium entering, in turn, is determined by the voltage-sensitive gates. When the gates are open, a set level of permeability exists and a constant amount of sodium ions diffuses in. Above threshold, any stimulus to the neuron, regardless of strength, has the same result — opening the gates. These gates always allow roughly the same amount of $Na^+$ in each time and then close at a constant potential, thus preventing influxes of $Na^+$ beyond that potential.

**Problem 9.3** The special sense organs are highly modified sensory receptors that initiate such very different sensations as sight, hearing, taste, smell, and balance, yet they transmit essentially identical nerve impulses. How can such different sensations be reconciled with such similar impulses?

The essentially identical impulses go to different regions of the brain for translation. It is thus the final destination of the impulses, not their intrinsic nature, that provides the diversity of sensation.

# Chapter 10
# MUSCULO-SKELETAL SYSTEM

IN THIS CHAPTER:

✔ *Introduction*
✔ *Invertebrate Support Systems*
✔ *The Vertebrate Endoskeleton*
✔ *Movement*
✔ *Muscle Types*
✔ *Solved Problems*

## Introduction

The bony skeleton is the chief structural system of vertebrates. It provides the basic form and shape of the organism. In addition to the mechanical function of support, the skeletal system also serves as the main means of protection for vulnerable organs within the body. In conjunction with the muscular system, the skeletal system allows body movement.

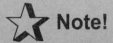

**Note!**

The skeletal system serves a homeostatic function in at least three ways: (1) locomotion as a behavioral response to environmental change, (2) production of erythrocytes in response to low oxygen levels in the tissues, and (3) storage of minerals such as calcium and phosphorus.

## Invertebrate Support Systems

**Hydrostatic Skeletons.** Support systems in early invertebrates depended on the relative incompressibility of water and related fluids. Such **hydrostatic systems** still extant consist of a fluid-filled cavity surrounded by muscle.

**Exoskeletons.** The **exoskeleton** of the arthropods provides structural support and protection for underlying soft parts and prevents drying out by means of a waxy outer layer found in terrestrial forms. Muscles are attached to different parts of the skeleton in an arrangement that permits movement around flexible joints. In the case of the exoskeleton, the muscles are attached to the inside of the skeleton.

## The Vertebrate Endoskeleton

Perhaps the most efficient support system exists in the **endoskeleton** of the vertebrates, an internal latticework of bone and cartilage which shapes and protects the body and provides a set of levers for movement. Rigidity of the internal skeleton of vertebrates is combined with a flexibility of the bony framework achieved by movement at their connections, or **joints**. The 206 individual bones of the adult human skeleton (see Fig. 10-1) are found in either the **axial skeleton**, consisting of the skull, vertebral column, and rib cage, or the **appendicular skeleton**, made up of the arms and legs and the girdle (pelvic and pectoral) to which these appendages are attached.

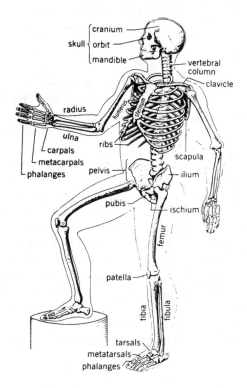

**Figure 10-1 The human skeleton.**

# Movement

Though other means of movement exist (amoeboid movement by **pseudopodia**, beating of cilia and flagella, etc.) the contraction of muscle is probably the most significant. The bones of the axial skeleton tend to provide maximum protection for the brain, spinal cord, and the soft organs of the chest. The separate bones of the skull are joined by **sutures**, which permit virtually no movement around that tight fit. The bones of the appendicular skeleton have much greater flexibility at their joints. These bones are held together by tough connective tissue bands, the **ligaments**, which permit a variety of movements at the joint. The muscles responsible for movement of the skeleton are attached to the bones by stringy connective tissue strands called **tendons**.

## Muscle Types

**Skeletal Muscle.** The muscle that moves the skeleton is known as **skeletal muscle**; it is also called voluntary, or **striated muscle**. This muscle is attached to bone, is under voluntary control, and contains a pattern of cross-sectional bands when viewed under the microscope.

The basic unit of the muscle is the muscle cell, or muscle fiber. However, skeletal muscle has several levels of organization both above and below this cellular level. Muscle fibers are grouped into bundles called **fasciculi**. These bundles form the muscle. The muscle fiber itself is a long, slender cell with multiple nuclei and a cell membrane called the **sarcolemma**. Within each fiber are many rodlike bundles called **myofibrils**. Each of these consists of a repeating pattern of structures that are seen in the myofibril as cross-sectional bands of different shading (see Fig. 10-2). These bands cause the striation seen in skeletal muscle fibers. A single repetition of this myofibril banding pattern is called a **sarcomere**, delineated by the **Z line**. Extending longitudinally are strands of a thin protein, **actin**. Toward the center of the sarcomere these actin strands interdigitate with another longitudinally aligned, thicker protein, **myosin**.

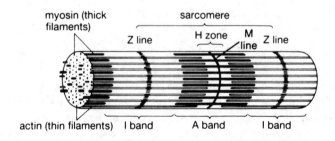

**Figure 10-2 Detail of a myofibril showing a complete sarcomere.**

94

## Remember

Z – denotes the end of the alphabet
and the ends of the sarcomere
M – for **Middle**
H – only thick filaments
I – only thin filaments
A – "**A**" little of both

Each sarcomere is surrounded by a dense network of ducts, the **sarcoplasmic reticulum**, which contains high concentrations of $Ca^{2+}$ ions that are released during muscle contraction. Invaginations of the sarcolemma along the Z lines form deeply penetrating tunnels called **transverse** or **T tubules**. These are involved in conducting action potentials throughout the fiber. According to the sliding filament theory, muscle contraction occurs through the repeated attachment of myosin myofilaments to actin filaments, followed by a pulling of the actin filaments from both ends of the sarcomere in toward the center, with a consequent shortening of the fiber. The two filament types do not contract, but rather simply slide past each other.

**Cardiac Muscle.** Cardiac muscle is primarily composed of striated, mononucleate cells. The junction of two such cells is characterized by an **intercalated disk**, a tightly fitting adaptation of the two membranes that offers minimal resistance to the passage of action potentials from one cell to the next. The branching of the cells and the ease of passage of action potentials between them enable depolarization to sweep across the heart and induce almost simultaneous contraction of the cells of a chamber. This unified contraction is necessary to generate sufficient pressure for pumping blood throughout the body.

**Smooth Muscle.** Smooth muscle fibers are mononucleate and much smaller than skeletal muscle fibers. They usually exist in tightly packed groups in organs such as the gut, uterus, and ureter. The same elements and processes of contraction operate in smooth muscle as in striated muscle, but in smooth muscle there appears to be much more actin and less myosin. Smooth muscle is also capable of maintaining contractions over long periods.

## Solved Problem

**Problem 10.1** During contraction, what changes would you expect to see in the various lines, bands, and zones of the sarcomere, on the basis of the sliding filament hypothesis?

During contraction, the actin filaments are drawn inward along the myosin filament. This causes the Z lines to be drawn closer together. Also, overlapping of actin and myosin increases; therefore, the I band, which is unoverlapped actin, gets smaller. The A band corresponds to the complete myosin myofilament, and since myosin does not change length during muscle contraction (nor does actin), the A band does not change. However, the H zone, which is unoverlapped myosin, gets smaller, and the darker margins of the A band get larger. The M line remains the same.

**Problem 10.2** Each skeletal muscle fiber, or rather each **motor unit** consisting of one or a number of muscle fibers, is innervated by a single axon. A stimulus sufficient to excite depolarizes the membranes of each muscle fiber making up the motor unit and produces an all-or-none response in each fiber, yet skeletal muscle is capable of graded responses of varying strengths. How is this possible?

An entire muscle consists of many motor units. Stronger stimuli involve a greater number of motor units. Thus, a graded response is in part dependent on the number of motor units responding to a particular stimulus.

The situation is somewhat complicated by the fact that axon stimulation usually occurs as a series of action potentials moving along the nerve rather than as a single impulse. These volleys of impulses may be summated for any single motor unit and thereby produce a greater response than would be the case for a single impulse. Thus, summation effects of a single unit also contribute to graded responses.

The existence in muscle of **fast** (quickly responding) and **slow** (more sluggishly responding) fibers also modifies the total contractile response to a stimulus.

# Chapter 11
# RESPIRATION AND CIRCULATION

IN THIS CHAPTER:

- ✔ Mechanisms of External Respiration
- ✔ Mammalian Respiration
- ✔ Exchange of $O_2$ and $CO_2$ in the Blood
- ✔ Circulatory Systems
- ✔ Cardiovascular Systems of Vertebrates
- ✔ The Human Heart
- ✔ Arteries, Veins, and Capillaries
- ✔ Blood Pressure
- ✔ Blood and Lymph
- ✔ Solved Problems

## Mechanisms of External Respiration

**External**, or **organismic**, **respiration** is concerned with providing a means to exchange gases necessary for ($O_2$) and as end products of ($CO_2$) cellular respiration. In unicellular and small, flat, multicellular organisms, gas exchange is accomplished quite readily across the cell membranes. The opposing passages of $O_2$ and $CO_2$ across the membrane are influenced by the partial pressure of the gases in the external environment.

One adaptation for bringing air more rapidly into the interior is the **tracheal system**, a system of tubes that ramifies throughout the body of the organism (such as spiders and insects) and carries air to individual cells. The larger tubes are known as **tracheae** and arise from apertures along the surface of the body called **spiracles**. The spiracles may open or shut in accordance with the action of valves. The large tracheae are maintained as open tubes by supporting rungs of **chitin**, a tough nitrogen-containing polysaccharide.

In most aquatic animals the respiratory organ consists of a series of protruding flaps known as **gills**. These gills are richly endowed with blood vessels engaging in the vital exchange of gases. In most bony fish, water enters the mouth cavity and is forced out through five sets of gill arches on each side of the head.

The lungs of the vertebrate represent one of the most vital evolutionary adaptations for life on dry land. By internalizing the moist respiratory surface participating in gas exchange with the surrounding air, lung breathers could modify their skins for other purposes.

## Mammalian Respiration

The breathing apparatus in mammals is exemplified by the human respiratory system (see Fig. 11-1).

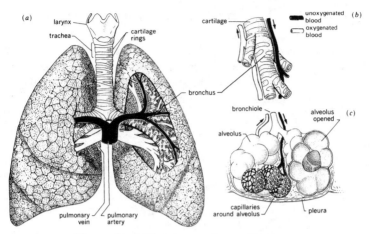

**Figure 11-1 Human respiratory system.**

Air may enter through either the nose or the mouth. Inhaled air passes through the **pharynx** and then into the **trachea**, a rigid tube resting in front of the esophagus. The opening of the trachea (the **glottis**) is guarded by the flaplike **epiglottis**. The **larynx**, or voice box, lies at the origin of the trachea. The trachea bifurcates into a left and a right **bronchus**. Each of these bronchi subdivides into many **bronchioles**, which constitute the small branches of the respiratory tree.

Each bronchiole terminates in a cluster of tiny air sacs called **alveoli**, which are heavily supplied with capillaries. The main events of external respiration occur across the alveolar membrane. The capillaries surrounding the alveoli bring $O_2$-depleted, $CO_2$-laden blood from the pulmonary arteries. The alveoli, on the other hand, upon inhalation contain a high partial pressure of oxygen ($pO_2$) and a low partial pressure of carbon dioxide ($pCO_2$). The resulting pressure gradients lead to rapid diffusion of these two gases across the capillary and alveolar walls → oxygen enters the capillaries and is carried to the tissues; carbon dioxide enters the lungs and is expelled upon exhalation.

The mammalian lung plays an essentially passive role in the breathing movements—inhalation and exhalation. The pulmonary cavity is alternately expanded and contracted by contraction of the **diaphragm** and **intercostal** muscles. The increase in cavity size draws air into the lungs from the outside as a result of the suction created. A decrease in size when the muscles relax pushes air out.

## Exchange of $O_2$ and $CO_2$ in the Blood

The $CO_2$ that enters the bloodstream from metabolically active tissues either dissolves in the plasma or enters the red blood cells, where some of it associates with the amino groups of the amino acids in **hemoglobin** (the main protein involved in $O_2$ transport) forming **carbaminohemoglobin** (HbNHCOOH). The remaining $CO_2$ combines in the red blood cells with water to form carbonic acid ($H_2CO_3$), which quickly dissociates into the bicarbonate ion ($HCO_3^-$) and $H^+$. The $HCO_3^-$ diffuses into the plasma.

When, the blood reaches the lungs, most of the $CO_2$ that was bound as carbaminohemoglobin or as the bicarbonate that diffused into the plasma is converted back to free $CO_2$ and diffuses into the lungs, where it is exhaled.

The **Bohr effect** is a decrease in affinity of hemoglobin for $O_2$ in the presence of high levels of $CO_2$ (or low pH). This effect facilitates the release of $O_2$ to tissues that have been metabolically active and consequently have little $O_2$ and high levels of $CO_2$.

**Important!**

The Bohr effect allows hemoglobin to easily deliver oxygen to the tissues that need it most!

In the oxygen-rich lungs, where uptake, rather than release, of $O_2$ is required, hemoglobin demonstrates another valuable property: **cooperativity**. When one $O_2$ molecule forms its noncovalent bond with one of the four heme groups, it alters the shape of the heme in such a way that the $O_2$ affinity of the other hemes increases. This cooperativity permits rapid uptake of $O_2$ in the lungs.

## Circulatory Systems

Cell volume is limited by the physical constraints simple diffusion places on gas exchange. In addition, as multicellular organisms become increasingly complex, cells buried in the interior are farther removed from the cell layer where gas exchange occurs and so become increasingly limited in their ability to obtain and eliminate gases through simple diffusion. Thus, various types of circulatory systems have evolved to supplement diffusion.

In all vertebrates and several groups of invertebrates (e.g., earthworms) the circulatory system is a completely closed system of tubes. Other organisms have an open circulatory system, in which the tubes are contiguous with open regions, or **sinuses**. Such open systems show a lowered transport efficiency and a slower circulatory time.

**Example 11.1** Insects possess an open circulatory system. As a result, their circulation of bloodborne materials, including $O_2$, is relatively slow compared with that of vertebrates. Yet, they have a very high level of metabolic activity. Such activity is possible because the oxygen necessary for combustion of their fuel is provided outside the circulatory system by a tracheal system of respiration.

## Cardiovascular Systems of Vertebrates

The cardiovascular system of all vertebrates consists of a muscular pump, the **heart**, and a system of tubes that carry blood to and from the heart. Vessels that carry blood away from the heart are called **arteries**; those that carry blood toward the heart are known as **veins**. Smaller arteries are called **arterioles**; smaller veins are designated **venules**. The connectors between arterioles and venules are the **capillary beds**, in which the actual exchanges between blood and tissues occur.

In primitive animals the heart is little more than an expanded section of blood vessel. In fish, the earliest of the vertebrates, the heart consists of a single thin-walled receiving compartment, the **atrium**, or **auricle**, which empties into a thicker-walled, more powerful pump chamber, the **ventricle**. Blood is pumped to the gill capillaries for oxygenation, but because the fish does not have a pump on the other side of the gills, the blood moves slowly and with considerably diminished force through the rest of the body (**systemic circulation**).

Amphibians have a left and a right atria, but a single ventricle. In the amphibian two circulations are evident, a pulmonary and a systemic circuit. The lungs in most amphibians are hollow and relatively inefficient; the skin acts as an adjunct oxygenation organ, and some of the blood that moves toward the lungs is actually shunted through the skin.

In both birds and mammals a clear separation of the two circulations is apparent, and the four-chambered heart can actually be viewed as two hearts. The right side of the heart receives deoxygenated blood from the systemic circulation and pumps it to the lungs through the pulmonary artery. Oxygenated blood returns to the heart through the two pulmonary veins and is pumped out the left ventricle into the aorta, from which it eventually reaches all parts of the body.

## The Human Heart

The human heart is a four-chambered structure located in the chest (see Fig. 11-2). The heavy muscular portion comprises the two ventricles; the two atria appear as flaps lying atop the ventricles.

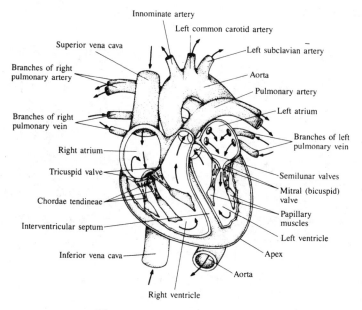

**Figure 11-2  The human heart.**

During filling, the atria are in a relaxed state. In response to an electrical impulse generated by the **sinoatrial** node (SA node, pacemaker), located in the right atrium, a wave of muscle contraction sweeps over the atria and blood is sent to the relaxed ventricles. The relaxed state of heart muscle is called **diastole**; the contracted state is known as **systole**. Systole in the atria is well over before the ventricles go into systole. This contraction of the ventricles would tend to send the blood back into the atria if not for the action of **atrioventricular** (AV) **heart valves**, which lie between each atrium and ventricle. These valves consist of flaps of connective tissue that prevent backflow of the blood. A wavelike spread of the impulse throughout both atria occurs, and these chambers undergo systole. A slowing down then occurs, as the impulse travels into the **atrioventricular node** (AV node). The spread of impulses along the fibers imposes a delay of over 0.1 s and thus assures that the atria have finished their systole before ventricular contraction begins. The impulse at the AV node then fans out through a bundle of fibers called the **AV bundle**, or the **bundle of His**. These fibers, called

**Purkinje fibers,** leave the AV bundle and carry the impulse rapidly throughout the ventricles.

# You Need to Know

### Flow of Blood:

vena cava → right atria → right ventricle → pulmonary artery → lungs → pulmonary vein → right atria → right ventricle → aorta → systemic circulation → vena cava . . .

## Arteries, Veins, and Capillaries

Arteries are thick-walled vessels that carry blood away from the heart. They consist of a smooth, elastic inner layer, a muscular middle layer, and an outer layer of fibrous connective tissue. In all arteries except the pulmonary artery the blood is oxygenated.

Veins lack the elasticity of the arteries; however, they are readily dilated by blood moving through them. Veins lack the pumping pressure of the heart to keep the blood flowing toward the heart; instead they rely on a series of one way valves working in concert with the squeezing pressure exerted by the routine activity of nearby skeletal muscles.

The significant exchanges between cells and the circulatory system occur in the **capillary beds**, networks of small tubules that stand between the afferent (entering) arteriole and the efferent (leaving) venule. Most of the capillaries are composed of a single layer of cells.

**Example 11.2** In a continuously flowing river, where the riverbed widens, the rate of flow diminishes. Where the riverbed narrows, the speed increases. The rate of flow then is inversely proportional to the cross-sectional area of the flowing column. The same relationship holds for the column of blood flowing through the vessels.

## Blood Pressure

**Blood pressure** refers to the force exerted on the blood vessels by the blood contained within them. During ventricular systole, arterial blood pressure is at its highest, and during diastole at its lowest. Average blood pressure for a young adult is 120 (systolic)/80 (diastolic).

---

### Remember

**Hypertension** is associated with a systolic above 150 or a diastolic above 95 mmHg. An elevated diastolic, particularly if it exceeds 100, is usually regarded as the more serious threat to health.

---

## Blood and Lymph

**Blood** is a fluid made up of liquid plasma, the primary component of which is water, and floating cells. The plasma is endowed with dissolved salts, proteins, lipids, and carbohydrates.

The cellular constituents of the blood, usually described as the **formed elements**, consist of three types:

1. Red blood cells, or **erythrocytes**
2. White blood cells, or **leukocytes**
3. **Platelets (thrombocytes)**

The pH of blood is maintained within a narrow range of 7.3 to 7.5 in healthy individuals. The maintenance of a constant pH is dependent on the action of buffer systems present in the plasma and in the red blood cells—mainly the carbonic acid/bicarbonate ion system.

In mammals, red blood cells are relatively small and, when mature, lack a nucleus and other organelles such as mitochondria, but are full of hemoglobin.

White blood cells are primarily involved in guarding against invading organisms that are often associated with disease.

Platelets are fragments of larger cells that function chiefly as initiating factors in blood clotting. The clotting, or **coagulation**, of blood involves the formation of a solid mesh of fibrils from soluble factors contained in the plasma through a cascade of biochemical events. The clot proper consists of a network of fibrin and blood cells.

**Lymph** is a yellowish fluid derived from the interstitial fluid with an osmotic potential considerably less than that of plasma but otherwise quite similar to it. It is carried in lymphatic circulation, along which **lymph nodes** (glandular tissue associated with the immune system) are located. The lymphatic system returns fluid and proteins to the blood.

## Solved Problems

**Problem 11.1** Why is the circulatory system so much more important for external respiration in the earthworm than in the insect?

The earthworm breathes through its skin, which is kept moist at all times through mucous secretions. Once $O_2$ has penetrated the skin by diffusion, it must be carried to all parts of the earthworm's interior. In the same way, $CO_2$ and $H_2O$ must be transported by the blood from the tissues where they are produced to the skin, where they may be discharged. The absence of an adjunct distribution system would not permit a sufficient rate of gas exchange to sustain metabolic needs. Therefore the earthworm requires a circulatory system.

The insect has a tracheal system. This ramified system of tiny, relatively rigid tubes that terminate in tiny pockets adjacent to the internal tissues enables the atmosphere to "reach into" every nook of the insect's body and obviates the need for further internal transport.

**Problem 11.2** Oxygen tensions in the alveoli, the pulmonay artery, and the pulmonary vein are all different. Match each of the following partial pressures with these three sites: 40 mmHg; 160 mmHg; 100 mmHg.

The pulmonary artery carries oxygenated blood to the lungs. We would therefore expect it to have the lowest oxygen tension of the three given: 40 mmHg. Since it is the source of the $O_2$ used by the body, inspired air of the alveoli has the highest of the three: 160 mmHg. By the time the blood has passed through the capillary beds around the alveoli and into the pulmonary vein, its $pO_2$ has been raised from 40 to 100 mmHg, which is th average $pO_2$ of the air in the alveoli.

# Chapter 12
# HOMEOSTASIS AND EXCRETION

IN THIS CHAPTER:

✔ *Homeostasis*
✔ *Excretion*
✔ *Invertebrate Systems*
✔ *Vertebrate Systems*
✔ *Homeostatic Function of the Kidney*
✔ *Solved Problems*

## Homeostasis

Living cells, as well as larger multicellular organisms, can function adequately only within a relatively narrow range of temperature, pH, ion concentration, sugar levels, etc. The maintenance of constancy of these conditions is called **homeostasis**. All the organ systems of the body operate cooperatively to effect internal constancy.

**Feedback Control.** The detection of changes in physiological state is mediated by stimulation of sensitive receptors. **Feedback** involves the creation of conditions such that any process A will lead to creation of B that regulates (feeds back on) the initial process A. If B inhibits A, then the control loop (circuit) is said to constitute **negative feedback**. If B increases A, then the control loop constitutes **positive feedback**.

**Regulation of Temperature. Homeothermy** is the capacity of certain groups of animals to maintain a constant body temperature despite fluctuations in the environment. Once called "cold-blooded", the **poikilotherms** actually demonstrate an equilibrium of their internal body temperature with the temperature of the external environment.

Homeostatic mechanisms for temperature control exist at many levels. Behavioral responses include avoidance of the source of heat, suppression of physical activity, and selection of approriate clothing.

In overheated conditions blood vessels at the skin's surface dilate, bringing an increased volume of blood to the surface where heat can be dissipated. If heat dissipation is not sufficient to maintain temperature constancy, the sweat glands secrete copious quantities of salty fluid (**sweat**). The vaporization of sweat results in a cooling effect. In some animals panting increases the exchange between moist membrane surfaces and the air, giving a cooling effect.

Many of the mechanisms for warming the body are reversals of the cooling processes: however, several are unique to the heating process. **Piloerection** is the standing up of individual hairs. During piloerection an increase in the thickness of the insulating layer is augmented by the trapping of stationary air within the matting of hair or fur. **Shivering** may also be invoked. Shivering is a spasmodic, relatively uncoordinated series of muscular contractions that produces a great deal of heat in a short period of time.

**Regulation of Blood Sugar.** Glucose is the primary source of carbohydrate fuel carried around in the blood. The careful regulation of blood sugar is an important aspect of homeostasis.

The hormone **insulin** reduces the levels of sugar in the blood by promoting the utilization, storage, and metabolic conversion of glucose stores. Insulin is extremely sensitive to blood sugar levels and is part of a negative feedback circuit that maintains constancy. As soon as blood sugar levels begin to rise, the $\gamma$ cells of the islet tissue of the pancreas increase both the synthesis and release of insulin. The effect of an increased insulin titer in the blood is a reduction of glucose concentration.

## You Need to Know

A failure to produce insulin, a lack of sufficient insulin, or a refractoriness to insulin's effects causes the disease **diabetes mellitus**.

A number of other hormones influence blood sugar levels directly or indirectly. **Epinephrine, glucocorticoids, thyroxin, somatotropin,** and **glucagon** all result in raised glucose levels in the blood. The regulation of blood sugar is interdependent of other carbohydrates or the pathways of lipid and protein metabolism.

# Excretion

**Excretion** is the process by which an organism rids itself of metabolic wastes. The major excretory product is $CO_2$, which arises from the degradation of organic fuel molecules. Most of it is channeled via the blood to the external environment. Another significant excretory product is nitrogen. When protein is utilized as a fuel, its amino acids must first undergo deamination or transamination. Eventually, the nitrogen removed is excreted as ammonia, urea, or uric acid. The final form of the nitrogen end product excreted is determined by the availability of water. If water is freely available, the simplest mode of nitrogen release is $NH_3$ formation. However, since this compound is quite toxic, the danger of its buildup has imposed evolutionary constraints upon some organisms. One way of neutralizing the $NH_3$ is to combine it with $CO_2$ to form **urea**.

Humans are **ureotelic** (the nitrogen excretory end product is urea). Urea is produced in our livers and brought to the kidneys for excretion in the urine. **Uricotelic** organisms produce a much more complex product known as **uric acid**. Uric acid is also nontoxic, but because it is relatively insoluble in water, far less fluid is needed to dispose of it. However, there is a higher energy cost in its formation.

Excretion is intimately tied to fluid and electrolyte homeostasis; many of the structures associated with waste disposal play a key role in water balance as well. Excretory organs such as the kidney not only process materials for elimination, but actively regulate the levels of many substances in the blood.

## Invertebrate Systems

Among single-celled Protista, and even the sponges, excretion is usually achieved by a **contractile vacuole**, which is a fluid-filled organelle that periodically contracts to force fluid, salts, and dissolved waste materials out of the cell. However, some protists lack these special organelles and expel their waste materials directly across the permeable cell membrane.

A clearly defined excretory system exists in the flatworms. It consists of two or more highly branched longitudinal tubules. Some side branches of the tubules terminate in a hollow bulb into which long cilia project. These cilia create continuous currents which carry fluid and waste materials out of the body through the **excretory pores**.

In the earthworm, which has a closed circulatory system, excretion is carried out by close association of the circulatory system with a unique set of **nephridia** (see Fig. 12-1). Body fluids enter a nephridium through the membrane of the **nephrostome**. The nephridium then gives rise to a coiled tubule, which is intimately associated with a blood capillary. This association of excretory structure and blood vessel, which allows reabsorption of material, is a functional forerunner of the vertebrate kidney. The nephridium terminates in a large bladder that opens to the exterior by means of a **nephridiopore**.

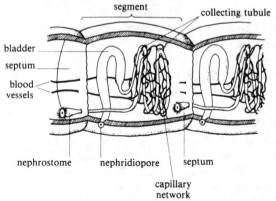

**Figure 12-1 Earthworm segment showing a nephridium.**

Insects have evolved a set of excretory structures called **Malpighian tubules**. These tubular sacs are washed by the blood of the insect and absorb fluid and waste materials at their closed ends. Uric acid is formed within the tubules, and fluid and salts are reabsorbed by the open blood system surrounding the tubules. The urine formed in the tubule moves into the hindgut and out of the body through the rectum. A great deal of water reabsorption occurs in the rectum, so that the feces and urine, both of which exit through the anus, are quite dry. This ability to conserve water in the excretory process is highly adaptive for organisms that have successfully invaded dry land.

# Vertebrate Systems

**Vertebrate Anatomy.** The kidney is an organ unique to the vertebrates. It is the chief excretory unit in higher vertebrates, but its major function in the lower vertebrates (fish) is osmoregulation. Nitrogenous wastes are handled largely through the gills of fish.

The **cloaca** is the chamber adjacent to the tail into which urine, sperm and eggs, and feces are deposited in their egress to the exterior. In mammals the cloaca is no longer present in the adult, and separate openings for digestive wastes and the urogenital system exist.

The two human kidneys are located laterally at the back of the abdominal cavity. On gross scale the kidney is divided into three regions: an outer **cortex**, a middle **medulla**, and an inner **pelvis** (see Fig. 12-2). Urine formed in the outer two layers collects in the renal pelvis and is transported to the urinary bladder by the **ureter** (one from each kidney). From the bladder the urine is transported to the exterior by the **urethra**.

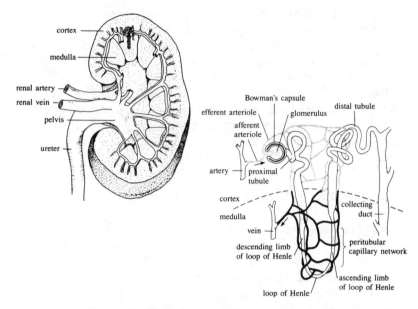

**Figure 12-2  The human kidney and nephron.**

The actual production of the urine occurs in the many **nephrons**, the functional units of the kidney (Fig. 12-2). Each nephron consists of three parts. (1) The **glomerulus** is a tight ball of capillaries that filters blood through its walls into the second component of the nephron. Because of the high hydrostatic pressure in the glomerulus, almost all components of the blood, except proteins and the formed elements, are squeezed out of the capillaries. (2) The filtrate passes into the second component of the nephron, the **convoluted tubule**, beginning at the **Bowman's capsule**, a sac that surrounds the glomerulus and receives the glomerular filtrate through its cell walls. The remainder of the tubule consists of a **proximal** section, a middle **loop of Henle**, and a **distal** section. It is from the convoluted tubule that most of the ions, molecules, and water of the filtrate are reabsorbed back unto the bloodstream. (3) The forming urine next passes into the **collecting tubule**, from which additional water may be reabsorbed before the tubule empties the urine into the pelvis of the kidney. The glomerulus, Bowman's capsule, and proximal and distal convoluted tubules are in the cortex; the loop of Henle and the collecting tubules are largely in the medulla.

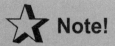

## Note!

The production of urine in the kidney involves three main processes:

1 **Filtration** = the nonselective passage of small molecules from the glomerular capillaries into the nephron
2 **Reabsorption** = essential small molecules are selectively returned to the interstitial fluid
3 **Secretion** = small molecules are selectively added to the filtrate for excretion

**Comparative Function.** Kidney function in vertebrates has evolved to adapt to unique living environments. In freshwater fish large amounts of fluid must be excreted, while salts must be conserved. No special devices for water reabsorption occur.

In marine fish the filtrate flows more slowly through the nephron, permitting a greater opportunity for reabsorbing fresh water from the filtrate. Marine fish drink salt water to overcome the loss of water to their hypertonic environment.

The habitat of most amphibians is generally rather moist, and their kidneys are quite similar to those of freshwater fish. Reptiles, however, are often found in dry environments, and their kidneys reflect the stresses of such a habitat.

Both mammals and birds possess kidneys that are remarkably efficient in terms of water conservation. The basic mechanism for producing a highly concentrated urine involves the creation of a markedly hypertonic environment around the nephron, so that water leaves the tubules by osmosis. Differences in the selective permeability of various regions of the nephron tubules also play a role.

The remarkable ability of the human kidney to produce a highly concentrated urine is believed to depend on the formation of a very steep concentration gradient in the renal medulla, through which a hypoosmotic urine must pass both in the descending arm of the loop of Henle and in the collecting tubule. A great deal of reabsorption from the crude filtrate is thought to occur in the proximal convoluted tubule. Water, amino acids, glucose, and ions all leave the tubule and are absorbed into the peritubular capillaries. In addition, certain substances such as excess hydrogen ions are secreted from the interstitial fluid into the filtrate as it passes through the nephron tubules.

## Did You Know?

**Sweating** is a form of excretion. Sweat, made in the sweat glands under the skin, is a source of abundant water loss, but it may also contain salts and even urea.

## Homeostatic Function of the Kidney

In addition to its excretory function, the kidney has a homeostatic function: it is a locus for homeostatic regulation of salt and water. Since most of the solutes in the plasma pass into the capsule of the nephron in the initial filtration process of urine formation, the blood needs to reabsorb from the rest of the tubule only those materials necessary to maintain appropriate homeostatic levels for each constituent. By regulating the amount reabsorbed, a balance in the blood may be effected at the renal level.

**Antidiuretic hormone** (ADH) is the primary agent for maintaining water balance. It is elaborated by a relatively small region of the **hypothalamus**, a highly versatile part of the brain lying just above, and helping to regulate, the pituitary gland. ADH increases the permeability of the walls of the collecting ducts to water. This retention of water results in the production of a concentrated urine. **Aldosterone**, a steroid of the adrenal cortex, regulates the active transport of sodium ions. The site of aldosterone action is the ascending arm of the loop of Henle. In the presence of high levels of the hormone, $Na^+$ is reabsorbed from the filtrate within the tubule into the surrounding vessels—a salt-conserving process.

# Solved Problem

**Problem 12.1** Describe one example of homeostatic feedback.

Osmoreceptors in the hypothalamic area of the brain provide a good example of homeostatic feedback. Under conditions of dehydration, the osmotic concentration of the blood increases. This solute-laden blood stimulates receptors in the hypothalamus, which have a low threshold to higher solute concentrations. In this particular circuit, the effector stimulated is a region of the hind lobe of the pituitary, which releases antidiuretic hormone (ADH). ADH acts on the kidneys, where it increases water reabsorption and concentrates the urine. Water retained in the body as a result of this will now oppose the dehydration and push the organisms back toward the original steady state.

**Problem 12.2** What are the advantages and disadvantages of eliminating nitrogenous wastes as ammonia?

When proteins are utilized as fuel molecules, or are otherwise bring broken down, their amino acids are deaminated, or stripped of their amino groups. This produces ammonia and an organic product that may enter the glycolytic pathway or be used for synthetic purposes. It would be simplest for the organism to deal with ammonia as the major nitrogenous excretory product, since it is the initial by-product and would thus require little additional energy to be excreted as is. However, a disadvantage of ammonia is its relatively high toxicity. This can be a serious problem where the water available is not sufficient to dilute the ammonia. For this reason, other nitrogenous end products have evolved. The formation of these alternative products requires both energy and organic materials.

# Chapter 13
# NUTRITION AND DIGESTION

IN THIS CHAPTER:

✔ *Food Procurement*
✔ *Digestive Systems*
✔ *Ingestion*
✔ *Digestive Enzymes*
✔ *Assimilation of Nutrients*
✔ *Egestion*
✔ *The Vertebrate Liver*
✔ *Vitamins*
✔ *Solved Problems*

## Food Procurement

All organisms require a steady supply of foods to provide fuel for their functional needs. Green plants, the common **autotrophs** (self-feeders), synthesize their food with the aid of sunlight from simple inorganic substances like $CO_2$ and $H_2O$. The fungi and nonphotosynthetic bacteria obtain their food by absorption from the immediate environment. The protozoans (a phylum within the kingdom *Protista*) and animals (multi-cellular heterotrophs) generally capture relatively large masses of food material.

Foods contain a variety of chemically defined **nutrients**, which provide materials for energy production and the structural substances needed for cell maintenance and growth. The major nutrients include carbohydrates, proteins, and lipids. Vitamins and minerals are also needed in smaller amounts.

Many animals live on vegetation alone; they are classified as **herbivores**. Other animals have a diet restricted to animal flesh; they are called **carnivores**. Still other animals eat both plants and animals; they are known as **omnivores**.

## Digestive Systems

Our discussion of digestion will be limited to the one-way digestive tract of those animals that possess a separate opening (**mouth**) for bringing food into the digestive tract and a second opening (**anus**) for ejecting the food remains. In the more primitive groups such as cnidarians and ctenophores (two-layered multi-cellular organisms) and the three-layered flatworms, a gastrovascular cavity is found that has a single opening to the environment.

## Ingestion

Mechanical grinding occurs in the mouth or further back along the digestive tract in a special chamber. In the case of some snails a sawlike device (**radula**) attached to the pharynx breaks the food down into smaller pieces. In earthworms, birds, and a number of other groups, a heavy muscular chamber (**gizzard**) behind the stomach or the crop grinds the food particles into smaller piece.

In vertebrates, teeth serve to grasp prey, tear bulk food, and grind it up. Carnivorous vertebrates usually have sharp teeth adapted for cutting and tearing. Herbivores have flatter teeth suited for grinding. Chewing food may accomplish some chemical, as well as the physical, breakup of food. This chemical degradation is achieved through the action of

**salivary glands**, which produce **saliva**. Its major function is wetting and lubrication of food particles. However, saliva also contains an **amylase**, which begins the breakup of starch that is ingested.

Considerable modification of the digestive tract is found in animals that do not capture and ingest large chunks or particles. Many insects subsist on fluids, such as blood. They generally have special piercing devices as components of their anterior ends.

The **filter feeders** obtain tiny food particles by straining large volumes of fluid from the environment through combs, small pores, or other seivelike structures.

In most animals the anterior opening of the digestive tract is followed by a **pharynx**, a muscular chamber that carries food to a tube called the **esophagus**. Food is moved through the esophagus by a sequence of contractions called **peristalsis** to a storage organ, the **crop** or **stomach**.

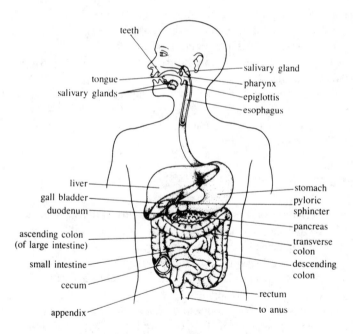

**Figure 13-1  Human digestive tract.**

**Example 13.1** A novel adaptation for grazing is found among the **ruminants**, large mammals such as cows that hastily swallow grass and weeds and then bring the material back up for further chewing and processing. The material brought back up is called the **cud**, and the rechewing helps to break down the cellulose. To accommodate the large mass of cellulose that is taken in, the ruminant has a four-chambered stomach. The first two chambers are actually expanded portions of the esophagus. It is from these two chambers that the cud is regurgitated. A huge population of bacteria and protozoans dwell in the rumen and reticulum, both of which act like fermentation vats in which the cellulose is slowly processed.

---

# Did You Know

A major waste product of fermentation is $CO_2$, which is eliminated during ruminant eructation . . . in other words, BURPING!

---

The next portion of the digestive tract is the long, tubular intestine. In all mammals the intestine is further divided into a long small intestine and a shorter large intestine. A great deal of chemical digestion and most of the absorption of nutrients into the circulatory system following digestion occur in the small intestine. A great length and highly folded surface of the intestine provides a correspondingly high absorptive surface area. Further, highly vascularized **villi**, finger-like projections of the mucosa, increase the contact between the nutrients moving and the boundary surface. In addition, each epithelial cell of the mucosal layer has tiny projections, the **microvilli**, along its surface to increase still further the absorptive lining.

## Digestive Enzymes

The mechanical breakdown of food, which occurs primarily in the mouth and stomach (or gizzard), is accompanied or followed by the chemical breakdown of nutrients by catalysts called digestive enzymes (see Table 13.1). These enzymes are chiefly involved in hydrolysis reactions:

$$\text{Polysaccharides } [C_6H_{10}O_5]_x + x(H_2O) \rightarrow x(C_6H_{12}O_6)$$
$$\text{Proteins} + H_2O \rightarrow \text{amino acids}$$
$$\text{Lipids} + H_2O \rightarrow \text{fatty acids} + \text{glycerol}$$

| SOURCE AND ENZYME | SUBSTRATE | SITE OF ACTION |
|---|---|---|
| Salivary Glands Ptyalin (amylase) | Starches | Mouth |
| Stomach Pepsin | Proteins | Stomach |
| Pancreas Amylase | Starches | Small intestine |
| Lipase | Fats | Small intestine |
| Trypsin | Polypeptides | Small intestine |
| Chymotrypsin | Polypeptides | Small intestine |
| Small intestine Peptidases | Peptides | Small intestine |
| Maltase | Maltose | Small intestine |
| Lactase | Lactose | Small intestine |
| Sucrase | Sucrose | Small intestine |
| Enterokinase | Trypsinogen | Small intestine |

**Table 13.1  Digestive enzymes.**

# Assimilation of Nutrients

The small intestine is the major site of digestion and absorption. A great deal of absorption of fluids and minerals also occur in the large intestine.

The monosaccharides, end products of carbohydrate digestion, are directly taken up by active transport into the circulatory system.

Facilitated diffusion and simple diffusion also are involved in the transport of simple sugars.

The absorption of the products of lipid digestion is quite complex. Smaller fatty acids may diffuse into capillaries and then be transported to the general circulation. Larger fatty acids may join other lipid materials to form complex lipid droplets known as **chylomicrons**, which eventually make it into the bloodstream. Many lipids move into mucosal cells as monoglycerides and diglycerides and may undergo changes in their degree of esterification intracellularly.

Amino acids and oligopeptides are transported into capillary beds of the intestine by active transport. In a few cases, passive diffusion may be the means of absorption.

These end products of digestion, which ultimately move into general circulation, are used as sources of fuel and for the synthesis of structural materials.

## Egestion

In humans, the small intestine joins the large intestine, or **colon**, which terminates in the **rectum**, where undigested material (**feces**) is collected before being passed out of the body through the **anus**.

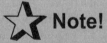 **Note!**

More than 60 percent of the fecal mass by weight consists of dead bacteria. Colon bacteria play a role in the absorption of water and minerals, produce certain vitamins, and maintain normal bowel activity. Digestive upsets caused by antibiotics are usually due to the destruction of intestinal bacteria by these antibiotics.

# The Vertebrate Liver

The vertebrate liver is both an **exocrine gland** (it sends secretions out through a localized duct) and an **endocrine gland** (it secretes substances directly into the bloodstream). These glandular functions are carried out in addition to the significant cycles of metabolic interconversions that take place. The liver aids in digestion by secreting **bile**, which emulsifies fats, aiding in their digestion and absorption. The liver produces albumin, cholesterol, and fibrinogen (a blood clotting factor) and stores iron and the fat-soluble vitamins A and D. It is the site for the conversion of hemoglobin to bile salts and bile pigments. The liver converts amino acids into small reserves of protein and also into glucose (during gluconeogenesis) under the driving force of glucocorticoids. The liver also strains the blood of particulate materials that are no longer useful, such as degenerating red blood cells. The liver also degrades many toxins in the body to harmless substances.

# Vitamins

Vitamins are special organic substances necessary in tiny amounts to sustain life. They usually function as coenzymes in a variety of metabolic reactions. Their absence usually causes a specific deficiency disease. Vitamins are usually identified by letters, but they also have chemical names. Some are water soluble (B complex, C and H) and others fat soluble (A, D, E and K). Table 13.2 lists a variety of vitamins and their functions.

| VITAMIN | FUNCTION |
|---|---|
| A | Integrity of epithelial tissues, eye function |
| D | Normal growth, bone formation |
| E or tocopherol | Antioxidative |
| K | Blood clotting |
| $B_1$, thiamine | Carbohydrate metabolism |
| $B_2$, riboflavin | Metabolism and electron transport system |
| Niacin or nicotinic acid | Dehydrogenation reactions |
| Folic acid | Growth, blood cell formation |
| $B_6$ or pyridoxine | Amino acid metabolism |
| Pantothenic acid | Carbohydrate and lipid metabolism |
| H or biotin | $CO_2$ fixation and fatty acid metabolism |
| $B_{12}$, cyanocobalamin | Blood cell formation, nucleic acid metabolism |
| C or ascorbic acid | Integrity of capillary walls, "intercellular cement" |

**Table 13.2  Vitamins.**

# Solved Problems

**Problem 13.1** List three advantages to an animal in having a large internal food storage capacity.

(1) The capacity to store food allows extra time for microorganisms to break down cellulose and thereby increases digestive efficiency; (2) storage capacity reduces the number of times that an animal must venture out in search of food and thus reduces the number of times it is vulnerable to predation; (3) without the ability to store food, animals in some cases would have to leave part of a kill behind, with a resultant waste of some of the energy that was put into making the kill.

**Problem 13.2** List four features of the intestine that increase absorptive capacity.

(1) Increased length, (2) extensive folding, (3) projection of finger-like villi into the intestinal lumen, (4) evagination of microvilli from epitheleal cells of the mucosa.

**Problem 13.3** Why do you suppose excessive doses of vitamins such as A and D pose a greater threat to health than vitamins such as C and the B complex vitamins do?

Vitamins A and D are fat soluble; because of this they move out of the bloodstream and accumulate in the body and thus are much more likely over time to reach harmful levels. Vitamin C and the B complex vitamins are water soluble and, so, are regularly excreted in the urine; they therefore do not build up so readily to harmful levels.

# Chapter 14
# REPRODUCTION AND EARLY HUMAN DEVELOPMENT

IN THIS CHAPTER:

✔ *Human Male Reproductive System*
✔ *Human Female Reproductive System*
✔ *Fertilization*
✔ *Embryogenesis*
✔ *Human Development*
✔ *Control of Differentiation*
✔ *Solved Problems*

## Human Male Reproductive System

The gonads of the male, within which the sperm and androgens are synthesized, are the **testes** (see Fig. 14-1). In many mammals the testes descend into an external sac, the **scrotum**. Sperm are formed in the **seminiferous tubules** of the testes. Scattered among the tubules are the **interstitial cells**, which continually secrete **testosterone**, the major androgen in the male. Androgens are secreted at high rates following **puberty**, the time at which sexual maturity occurs. Sperm are produced continuously and are stored in the **epididymis**, a highly coiled tubular structure lying atop each testis. The sperm are carried through the abdominal cavity by the **vas deferens**, a long tube that joins the **urethra** in the region of the **prostate gland**. The **urethra**, a tube that exits through the penis, carries urine and **seminal fluid** to the body's exterior. Sperm is carried in a fluid formed by the **seminal vesicles**, prostate gland, and **Cowper's glands**, accessory glands of the male reproductive tract. The penis delivers sperm (**semen**) to the female reproductive tract, but it must first become **erect**, or stiffened. Excitation increases and culminates in the expulsion of the semen (**ejaculation**) from the penis at climax of orgasm.

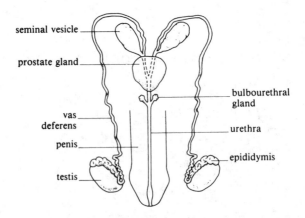

**Figure 14-1 Human male reproductive system.**

## Human Female Reproductive System

**Anatomy.** The gonads of the female are the paired ovaries found in the abdominal cavity (see Fig. 14-2). Overhanging each ovary are the **fallopian tubes** (**oviducts**), which lead to the muscular uterus. Eggs that are fertilized in the fallopian tubes are swept along by cilia into the uterus, where early development and eventual implantation will occur.

The uterus adjoins a short corrugated tube known as the **vagina**. The vagina stretches over the narrow neck of the uterus (**cervix**) at one end and extends to the outside at its distal end. The external sex organs—the **labia majora** and **minora** and the **clitoris** — that surround both the opening into the vagina and the separate opening into the urethra form the **vulva**.

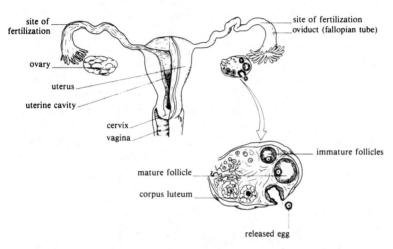

**Figure 14-2  Human female reproductive system.**

**Menstrual Cycle.** The **menstrual cycle** refers to the series of events that periodically modifies the female reproductive tract of humans and advanced primates. Other mammals have a cycle of receptivity to sexual intercourse known as the **estrous cycle**. The menstrual cycle is regulated by hormones produced by the hypothalamus of the brain, pituitary gland, and endocrine structures present in the ovary. **Gonadotropin-releasing hormone** (GnRH) from the hypothalamus of the brain stimulates the pituitary to produce two hormones, **follicle-stimulating hormone** (FSH) and **luteinizing hormone** (LH), which stimulate the ovaries to produce the female sex hormones **estrogen** and **progesterone**.

The total cycle may be divided into three parts: the **menses** (flow of blood), which lasts about 5 days; the **follicular** (or **proliferative**) phase, which follows the cessation of flow and lasts about 9 days; and the **luteal** phase, which comprises about the last 2 weeks of the cycle. During the 5-day period of flow, the levels of GnRH, FSH, and LH are all relatively low. Both estrogen and progesterone levels are also low, and the **endometrium**, or inner lining of the uterus, is being shed.

During the proliferative phase, **follicles** begin to develop in the ovaries under the influence of FSH and LH from the pituitary. Hypothalamic production of GnRH is appreciable. A single growing follicle continues to produce increasing amounts of estrogen, which influences growth and vascularization of the endometrium. At the time of **ovulation** (release of the ovum from the follicle) there is a sharp drop in estrogen and a surge in LH levels. This marks the transition from the follicular to the luteal phase and causes the conversion of the ruptured follicle to the **corpus luteum**. The corpus luteum secretes high levels of progesterone, which is responsible for the changes associated with the luteal phase. The thickened endometrium becomes highly glandular, and an increase in glycogen occurs. These maturational changes of the uterus prepare it for implantation and pregnancy. Very little GnRH is produced during this third phase of the cycle. One important effect of the high levels of progesterone is the suppression of follicular development in the ovary.

---

## Remember

It is progesterone that acts as the major ingredient of the birth control pill, since a woman is maintained in postovulatory condition under its influence, with no follicle formation.

---

Should pregnancy fail to occur, the corpus luteum will break down toward the latter part of the luteal phase (about day 21) and soon the endometrial lining will degenerate and a new menstrual flow will begin. The boundary between luteal and flow phases is marked by a sharp drop in progesterone. All other hormones are also low. However, with the loss of the corpus luteum a new buildup of GnRH occurs, and by the end of flow a buildup of estrogen will be renewed by newly developing follicles.

Menstruation ceases during pregnancy and trails off when older women enter **menopause**, the period of slackening of hormone secretions that control the menstrual cycle.

# Fertilization

The overwhelming advantage of sexual reproduction lies in its capacity to provide genetic variation. The obvious disadvantages of sexual reproduction is that, at least in dioecious forms, it requires a meeting of two different organisms.

External fertilization in aqueous surroundings guarantees transport and hydration of gametes. Consequently it is seen in aquatic animals and in amphibians. Other organisms, which have adapted to terrestrial habitats provide an internal medium for fertilization. In reptiles and birds, sperm are deposited within the reproductive tract of the female during coitus, and fertilization results in a zygote that is covered by albumin and a shell and is soon deposited externally to complete its development.

In the human female, fertilization actually occurs in the oviduct. Approximately a week later, the embryo implants in the uterus.

# Embryogenesis

In multicellular animals life generally begins with the union of the sperm and egg, resulting in a **zygote**. This activates the completion of meiosis in the egg and the sperm and egg pronuclei come together. Following fertilization the zygotic cell begins a regulated series of divisions called **cleavage**. Further divisions produce a cluster of cells called a **morula**. This is shortly followed by the **blastula** stage – where a thin layer of cells surround a central cavity, or **blastocoel**. In the next stage a series of cellular migrations and rearrangements occur which transforms the spherelike blastula into a double-layered cup, the **gastrula**. The outer layer of the gastrula will become the **ectoderm**. The inner layer of cells form the **endoderm**. The opening into the new cavity enclosed by the endoderm (the **archenteron**) is the **blastopore**. Later, an opening will form opposite the blastopore site. In echinoderms as well as in vertebrates, the blastopore becomes the anus, and the opposite opening gives rise to the mouth. In other multicellular animals the blastopore becomes the mouth. Concurrent with or slightly after early gastrulation two evaginations of the endoderm pinch off into the blastocoel and give rise to a third layer of cells known as **mesoderm**. The organs and organ systems of multicellular forms arise from undifferentiated cells of the primary germ layers through a process of increasing **differentiation**.

## You Need to Know

The three primary germ layers will soon produce a variety of differentiated structures:

*Endoderm* → digestive and respiratory tract lining, liver and pancreatic cells, lining of bladder and urethra, thyroid and pituitary glands.

*Mesoderm* → muscles, bone, cartilage, fiber, blood and vessels, mesentery, dermis, kidneys, reproductive organs

*Ectoderm* → outer skin layer, hair and nails, nervous system

## Human Development

The human egg is fertilized in the fallopian tube of the female. About 5 days after fertilization, a **blastocyst** is formed. This consists of an outer circle of cells known as the **trophoblast** and an inner round mass of cells hanging down at one end. The inner mass of cells will produce the embryo, while the trophoblast will give rise to the chorion and the fetal component of the spongy **placenta**. At about the seventh day the trophoblast buries itself into the endometrium. Soon after this **implantation** the extraembryonic membranes form. The amnion develops as a fluidfilled chamber within which the embryo proper is safely nestled. The chorion also produces human chorionic gonadotropin (hCG), which initiates the production of maternal estrogen and progesterone to support the pregnancy. By the second trimester of pregnancy, hormone production is a function of the placenta.

Pregnancy, or **gestation**, is divided into three parts. During the first trimester all the major systems have been formed. By the second week, mesoderm is formed. The heart and nerve tube have formed by the end of the third week. After two months the embryo takes on a human appearance and graduates to the status of **fetus**. During the second trimester, movements and reflex actions are quite pronounced and refinements in organ development occur. The major events of the third trimester are increases in body mass and size. After about 270 days of gestation, **parturition**, or birth, occurs.

# Control of Differentiation

The cells that make up a differentiating embryo are mostly genetically identical. Thus, the specialization process must involve the differential expression of genes. This may include transcription of some genes while other are suppressed, an increase in the activity of a particular gene, or the modification of post-transcriptional products. Cytoplasmic factors, already present in the egg and passed along to the blastomeres, during cleavage, may also be involved in embryogenesis.

The egg cell is capable of forming a complete organism. Such a capacity, which is shared with the blastomeres produced by at least the first three cleavages in mammals and amphibians, is termed **totipotency**. Beyond the early cleavage stages, most cells produced lose their plasticity and become destined to become specific structures within the embryo. Primitive structures including an **organizer**, orchestrates the development of more elaborate structures by secreting a small diffusive **inducer** molecule.

## Solved Problems

**Problem 14.1** Describe some asexual processes occuring in animal reproduction. What is a chief disadvantage of this form of reproduction? A chief advantage?

Male ants and bees develop via **parthenogenesis** (development of unfertilized eggs). Artificial stimulation of sea urchin eggs induces cleavage. Although the egg is a sex cell, the development of unfertilized eggs is an asexual process. **Budding** occurs among sponges and hydra. This involves the outgrowth of a portion of the parent's body to produce a new individual. In the phylum *Platyhelminthes* a process called **fragmentation** or spontaneous separation occurs. Each of the fragments produces a new flatworm. Related to fragmentation is **regeneration**, or restoration of lost parts.

Asexual reproduction is a simple procedure for producing progeny, but it tends to minimize the variation which is grist for the mill of evolution. In almost all animal organisms asexual reproduction is only a supplement to sexual reproduction.

An advantage of asexual reproduction is that an organism can quickly reproduce itself and establish a population in a particular environment.

**Problem 14.2** How do menstrual and estrous cycles differ?

Animals in heat experience a brief, strong sexual urge during the middle of estrous cycle but are not sexually receptive at any other time; sexual receptivity occurs throughout the menstrual cycle. Physically, estrus prepares the female reproductive tract for copulation, whereas the menstrual cycle involves an elaborate preparation of the endometrium for implantation of a fertilized egg. Consequently, if fertilization does not occur, any preparatory thickening of the uterine wall in estrous animals is merely reabsorbed; in menstruating animals the hypertrophic lining is sloughed off at menstrual flow. Finally, the events of the estrous cycle are much more subject to environmental influence than those of the menstrual cycle.

# Chapter 15
# EVOLUTION AND THE ORIGIN OF LIFE

IN THIS CHAPTER:

✔ *Introduction*
✔ *Darwin and Natural Selection*
✔ *Hardy-Weinberg Equilibrium*
✔ *Punctuated Equilibrium*
✔ *Speciation*
✔ *Origin of Life*
✔ *Solved Problem*

## Introduction

**Evolution** is the view that all reality is in a continual state of change. As applied to biology, evolution maintains that all the diverse forms of life have come into being through a gradual and continual process of modification of ancestral forms. This process of "descent with modification" does not lead to a finished final product. Evolution modifies all living things and will continue to produce change in the future as it has in the present and past. Though called a theory, evolution is accepted by many biologists as a fact.

133

## Darwin and Natural Selection

In the mid-1800s, two young naturalists, Charles Darwin and Alfred Wallace, independently developed essentially identical theories for **natural selection**, the proposed mechanism for evolution. However, Darwin is most closely associated with evolution for two significant reasons. First, he amassed such a comprehensive and convincing body of evidence demonstrating organic evolution that it was no longer reasonable to dispute the existence of such a process. Second, his explorations of the fauna of South America and Africa  during his five-year (1831–1836) voyage aboard the HMS *Beagle* provided him with the insights necessary to develop such a compelling theory. The theory of natural selection was first presented at a scientific meeting in 1858. In 1859, Darwin's landmark work, *The Origin of Species by Means of Natural Selection*, was published in London. It created a storm of controversy but generated a fervent band of supporters.

The theory of natural selection rests upon three major tenets. First, there are more offspring produced in a generation than can be supported by the limited resources (food, water, shelter, mates) of the environment. Second, heritable variations exist within this too large population of young. Third, a competition for survival occurs in which those variants that are better adapted to a particular environment are successful and continue to produce offspring with their adaptive characteristics. Over time the characteristics that confer adaptiveness, or **fitness**, come to accumulate in the population, while those characteristics that diminish fitness tend to dwindle or die out. It is this last aspect, the greater reproductive success of better adapted forms, that is properly termed natural selection.

## Hardy-Weinberg Equilibrium

The totality of the alleles of every gene in a population is the **gene pool** of that population. Changes in the frequencies of particular alleles constitute the raw material of evolution. Over long periods of time alterations in allelic frequency produce marked alterations in the characteristics of a population.

G. H. Hardy and W. Weinberg determined that the frequencies of alleles and even the ratios of genotypes tend to remain constant from one generation to the next in sexually reproducing populations under certain conditions. These conditions include:

1. A very large population
2. No change in mutation rates
3. Complete randomness in mating so that reproductive success is the same for all allelic combinations
4. No large scale migrations into or out of the mating pool

Think about it . . . how would failing to meet any of these conditions influence allele frequencies?

In such stable populations, gene frequencies follow simple laws of probability. For example, if allele A has a frequency of $p$ in a population, and allele B has a frequency of $q$, and there are no other alleles for this gene, $p + q = 1$. The probability that allele A will occur is equal to its frequency $p$; likewise, the probability that B will occur is $q$. Thus, in a given population the frequency of homozygous AA individuals is equal to $p$ x $p$, or $p^2$. By similar reasoning, the frequency of BB homozygotes is $q^2$. Since there are two ways of forming the heterozygote AB (the A allele from the mother and the B allele from the father, or vice versa), the frequency of AB in the population is $2pq$. The sum of all three genotype frequencies = $p^2 + 2pq + q^2 = 1$.

**Example 15.1** In a population in Hardy-Weinberg equilibrium, with only two alleles for a particular gene, if we know that allele A has a frequency $p$ of 0.3, we can find the frequency of allele B. Since $p + q = 1$, we know that $q = 1$ $p = 0.7$. Furthermore, we can determine the frequencies of the various genotypes, as follows:

$AA = p^2 = (0.3)(0.3) = 0.09$
$AB = 2pq = 2(0.3)(0.7) = 0.42 \quad BB = q^2 = (0.7)(0.7) = 0.49$

Since allelic frequencies remain constant from generation to generation, departures from this consistency help to expose selection pressures operating in the population.

## Punctuated Equilibrium

Students of evolution have recognized the comparative rarity of transitional forms in the fossil record between species or major groups such as reptiles and mammals. If all populations undergo slow changes in their evolution into new forms, one would expect to find a continuous spectrum of fossil representatives all through the transition process. The theory of **punctuated equilibrium** explained these gaps in the fossil record. In punctuated equilibria, evolutionary changes occur in irregular jumps, and very rapid changes are followed by long periods of relative constancy. Under certain conditions new species form from old, and the major modifications are compressed into several thousand years rather than many millions.

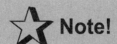 **Note!**

The theory of punctuated equilibrium has excited considerable controversy.

## Speciation

A **species** consists of all individuals who can share a common gene pool. This means that they may interbreed with one another to produce *fertile* offspring but not with members of another species.

**Example 15.2** The horse and the donkey (ass) belong to separate species. They do not share a common gene pool although they are capable of mating with one another to produce offspring. The cross of a male ass with a female horse yields a **mule**. A male horse crossed with a female ass produces a **hinny**. But the gene pools of each do not blend because both mules and hinnies are sterile and represent a reproductive dead end. Over time then, the horse and the ass maintain separate gene pools, the major criterion of species integrity.

Since species are separate from one another on the basis of their **reproductive isolation**, the key to maintaining species discreteness lies in mechanisms that produce separate breeding populations. One form of reproductive barrier is found in the **geographical boundaries** that prevent individuals from two populations from reaching one another. Alternatively, **biological mechanisms** may arise to maintain separateness between populations. These may include ecological, temporal, anatomical, or even behavioral differences.

In many cases in which sexual union does occur between separate species, the gametes do not fuse or an embryo fails to develop. Such **gametic incompatibility** is an effective barrier to a breakdown of species lines and if mating is successful, the hybrid forms are quite often sterile.

**Speciation** involves isolation of a segment of a species from other populations of that species. If the isolate group is separated *physically* from the original larger population, this type of speciation is called **allopatric**. **Sympatric** speciation is when a subpopulation within the parent group develops characteristics that isolate it from its neighbors, and it goes its merry developmental way to yield ever-increasing uniqueness while dwelling in the same neighborhood. **Adaptive radiation** occurs when a common ancestor reaches an environment in which a great number of distinct opportunities and challenges exist and result in the formation of many new species.

# Origin of Life

Biologists support the concept of a natural origin for life: that life emerges as a selected point along a continuous spectrum of increasingly complex arrangements of matter. When matter becomes sufficiently complex, we encounter the characteristics associated with life.

The hypothesis presented by A. I. Oparin in 1924 starts with the origin of the earth, about 4.6 billion years ago. The early atmosphere was a reducing one, possibly with large amounts of methane ($CH_4$), steam ($H_2O$), ammonia ($NH_4$), and some hydrogen ($H_2$). The environment favored the chemical synthesis of a variety of simple organic substances in the atmosphere, and these substances soon collected in the early seas.

The organic material of the seas, becoming increasingly concentrated, accreted into larger more complex molecules, special colloids termed **coacervates**. They tended to be shaped into droplets by surrounding "cages" of highly ordered water molecules. There thus existed a very clear line of demarcation between the molecules of the coacervate and the surrounding water. The absorptive properties of the droplet caused it to grow, and eventually an actual membrane may have formed at the coacervate-water boundary, increasing the selective permeability of the droplet.

Stanley Miller provided experimental support for the belief that the conditions and simple inorganic molecules present during the earth's early history could combine to create the complex organic molecules of living organisms. Miller, a student of the Nobel laureate Harold Urey, set up a Tesla coil that discharged electric bolts into a closed system containing methane, ammonia, water vapor, and some hydrogen gas. The results of this energetic stimulation of an atmosphere resembling that of the early earth were spectacular. A variety of organic molecules were generated, including ketones, aldehydes, and acids, but most important of all—amino acids. Since proteins are vital to both the structure and the function of living cells, the creation of amino acids under conditions that were believed to prevail upon the early earth supported the Oparin hypothesis.

It is now believed that carbon monoxide, nitrogen, and carbon dioxide were also significant constituents of that atmosphere. The Miller-Urey experiments of the 1950s were repeated using the revised atmosphere, and similar yields of organic molecules were achieved. This supported the earlier theories of a primordial transformation of inorganic molecules into the organic building blocks of life. Later it was shown that short polypeptides, riboses, and nucleotides were also produced.

## You Need to Know

The first organisms must have ingested pre-formed organic fuels for energy—that is, they must have been **heterotrophic**. However, it would have been evolutionarily advantageous to have the ability to synthesize nutrients from scratch, eventually giving rise to **autotrophs**.

## Solved Problem

**Problem 15.1** In order for organic molecules to form from the chemicals of the early earth's atmosphere, high sources of energy would have been necessary. What are some possibilities for these early energy sources?

Electric energy would have been abundant because of the prevalence of thunderstorms during the earth's early history. The numerous volcanoes provided a ready source of thermal energy. Radioactive energy was also abundant. Finally, because an ozone layer had not developed at that point, ultraviolet radiation would have been ubiquitous.

# Chapter 16
# ECOLOGY

IN THIS CHAPTER:

✔ *Introduction*
✔ *Energy Flow*
✔ *Nutrient Cycling*
✔ *Population Size*
✔ *Types of Ecosystems*
✔ *Ecological Succession*
✔ *Biomass*
✔ *Pollution*
✔ *Solved Problems*

## Introduction

**Ecology** is a holistic (broadbased and integrative) approach to understanding living things in context as they relate both to their physical environment (**abiotic** aspects) and to each other (**biotic** aspects). The ecological unit is the **ecosystem**, which is a group of diverse interacting populations found within the regional limits of a neighborhood. This neighborhood (**habitat**) might be as small as a local pond or as broad as the vast Sahara Desert. Various interacting populations within an ecosystem make up the **community**, the living components of the ecosystem.

## Energy Flow

Energy moves through the community of an ecosystem in a single direction by means of a **food chain** (web). Populations are assigned a **niche**, or occupational role in an ecosystem usually in terms of their relation to the overall flow of this energy (food) in the food chain:

**Producers**: the first group in a food chain, usually consisting of green plants, which convert some of the energy of the sun (through photosynthesis) into organic molecules they use and store in their tissues.

**Consumers**: animals that feed on green plants and each other. Primary consumers are herbivores, which subsist on the primary producing plants. Secondary consumers feed on primary consumers, while tertiary, quaternary, etc., consumers are further along the chain.

**Decomposers**: bacteria, fungi, plants, or animals that feed on dead organisms and release the bound organic material of the organisms to the food chain.

## Nutrient Cycling

Ecologists follow specific atoms through cycles, e.g., carbon (C), nitrogen (N), and sulfur (S), and chart their fates as they pass through the food chain, into the environment, and back again into the community. The nitrogen and carbon cycles are particularly well documented. We studied the nitrogen cycle in Chap. 8. The carbon cycle is powered alternately by the reduction (through photosynthesis) and the oxidation of carbon. Carbon enters the cycle as atmospheric $CO_2$, which is converted by plants to organic molecules during photosynthesis. Ultimately, this carbon is converted back to $CO_2$ through either respiration or combustion, and the cycle begins again.

## Population Size

Natural populations increase exponentially. A constant rate (reproductive potential) produces dramatic increases over time because as the population increases, the rate is being multiplied by an ever increasing base value (the number of individuals in the population). Growth increases slowly at first and then rapidly as the steady rate is applied to increasing numbers of individuals. But such a depiction is only hypothetical, since a population that increases in size at such a great rate will be subject to limiting constraints.

## You Should Know

The control of population size is one of several areas in which evolution and ecology coincide. Field experimentation in a variety of habitats has enabled ecologists to sharpen our understanding of the mechanisms of natural selection.

## Types of Ecosystems

In one sense, the ocean, land, air, and fresh water is a single ecosystem—the **biosphere**. Within the biosphere a traditional subdivision of major ecological types has been established. Land regions have been more intensively probed than marine or fresh aquatic habitats, but this does not reflect assigned importance or even intrinsic interest or complexity.

A **biome** is any of several unique terrestrial ecosystem types. The biomes constitute the largest community units classified by ecologists. The significant biomes and their characteristics are as follows:
1. **Tropical rainforest**: dense forests, warm temperatures, very heavy rainfalls, poor soil quality
2. **Desert**: extremely scanty rainfall, modest plant life
3. **Chaparral**: prolonged, hot, dry summers, temperate rainy winters, dominant vegetation forms are small trees and shrubs, small, bland colored animals
4. **Savannas**: tropical grasslands, light and seasonal rain
5. **Temperate grasslands**: limited water availability, clumps of scrub

grass, shrubs, and some annual plants predominate, small rodents coexist with large carnivores

6. **Taiga**: northern forests thick with massive conebearing evergreen trees, smaller animals such as hares, mice, shrews, and lynxes, and larger ones, such as bears, elks, deer, and moose, snow present most of the year

7. **Tundra**: northern grassland, permanent layer of frozen undersoil (permafrost) exists, short growing season during the northern summer provides sustenance for shrubs and rushes and for animal life (fauna), which includes multitudinous insects, birds, lemmings, and foxes

8. **Temperate deciduous forest**: trees that shed their leaves during the cold season, bushes and shrubs, and grasses interspersed with cryptogamic plants (mosses and liverworts), cold winters alternate with warm summers of adequate rainfall, animal life is abundant, ranging from mice, chipmunks, and raccoons to wolves and mountain lions.

## Try This . . .

Think of a representative geographical location for each biome.

Most of our planet's surface consists of water. The marine (saltwater) environment makes up about 70% of that surface. Both freshwater and marine environments possess a rich array of community life and significantly affect economic aspects of human societies.

From an ecological perspective, the oceans can be divided into a **neritic** region above the continental shelf and the **oceanic depths** beyond the relatively shallow shelf (see Fig. 16-1). The portion of the neritic region just offshore is named the **littoral zone**. Because of its currents and complete penetration by the sun due to its shallowness, it is particularly rich in plant and animal life. Shoreward of the littoral zone, an intertidal zone is periodically covered with water at high tide and exposed at low tide. The ocean depths are divided into a **pelagic zone**, rich in plankton, and the even deeper **abyssal zone**.

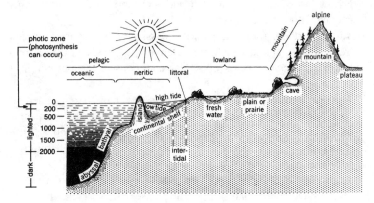

**Figure 16-1 Some common ecological environments.**

# Ecological Succession

Stability of ecosystems was once equated with complexity, particularly the complex interactions occurring within a food web. More recent field studies, however, show that some simple ecosystems may possess considerable durability; however, if only one population exists at a particular level, the loss of that population may doom the entire ecosystem. Although ecosystems demonstrate some flexibility and tend to maintain their integrity, they may be irreparably harmed by

1. Sudden shifts in the environment (temperature change, drought, flooding) that destroy a significant portion of the community
2. Uncontrolled increase in the numbers of particular populations due to failure of the mechanisms for population control
3. Loss of key minerals or other nutrients in the ecosystem
4. Human interference, which may lead to destruction of habitats, an overkilling of specific species, or pollution with toxic materials that cannot be handled within the ecosystem

Ecosystems evolve new characteristics and gradually supplant older communities with new populations. This slow change in the makeup of the community within a habitat is called **succession**. Succession often occurs as older inhabitants modify their environments to provide new opportunities for the next generation's plants and animals. Succession continues until a **climax community** is formed, one that is extremely well suited to the environment and remains essentially unchanged through long periods of time.

## Biomass

**Biomass** refers to the weight of living organisms within an ecosystem. Because of the continual loss of biomass in proceeding along a food chain the community can be considered a **pyramid** (see Fig. 16-2). In a food (energy) pyramid a broad producer base is topped by ever diminishing populations of consumers. The terminal consumer forms the apex of the pyramid.

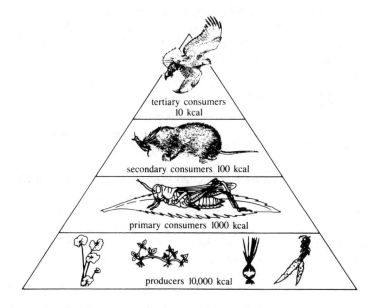

**Figure 16-2 Energy transfer in a typical food pyramid, starting with 1 million kcal of sunlight.**

## Pollution

**Pollution** is generally defined as the introduction of harmful materials to an ecosystem. Although pollution has been regarded as a human activity in which toxins and nondecomposing materials are brought into the flow channels of the ecosystem, it may also involve natural processes. Volcanoes and forest fires spew noxious ash and other atmospheric pollutants that can seriously damage or destroy ecosystems. One problem with pollutants, especially applicable to toxic organic substances, is that they become more concentrated as they move along a food chain. This is termed **biomagnification** or **bioaccumulation**.

**Eutrophication** involves too much of a good thing. An overabundance of nutrients is provided in the waters of a river or lake, stimulating overgrowth of **phytoplankton** (floating microscopic plants) or algae. This floral population soon reaches a density at which vital gases and nutrients are used up, and the overgrown "blooms" produce toxins and die as an unpleasant rotting mass.

## Solved Problems

**Problem 16.1** The biomass of primary consumers in a food chain is considerably less than the biomass of (primary) producers. Why?

All of the energy trapped within the plants of the first trophic level cannot be transferred intact to the next trophic level because of the waste and increase in entropy associated with each transfer of energy. In addition, the plant utilizes a great deal of energy it obtains from the sun to build up and maintain itself during its lifetime. Less than 10–15 percent of the calories stored by the plants pass into the herbivores that constitute the next trophic level.

**Problem 16.2** Explain how DDT can be found in only low concentration in the general environment yet occur at lethal levels in the inhabitants of the same area.

Like many toxic substances, DDT accumulates in fat tissue. Although it may initially be dispersed throughout the environment, the *producers* remove it from the general environment and concentrate it.

Thus, the primary *consumers* are eating higher concentrations than are found in the general environment. They too accumulate the toxins, and when eaten pass on even higher concentrations. Therefore, each trophic level exhibits toxic concentrations greater than those of the level below it.

**Problem 16.3** Organisms in an ecosystem interact in many different ways. One very common relationship is that of predator to prey. Another type involves long-term, intimate association between two different species: **symbiosis**. Describe three types of symbiotic relationships, differentiated by how the relationship affects both organisms. Give an example of each.

**Commensalism** is a relationship in which one partner benefits while the other is neither harmed nor benefited. An example is the remora, a fish that attaches itself to the underside of sharks by means of a sucker device atop its head. The shark, because of its constant activity and sloppy table manners, provides both transportation and "crumbs" for the remora. The remora does not advance the interests of the shark.

**Mutualism** is an association in which both members benefit. A lichen consists of a fungus intertwined with a green alga. Lichens are particularly prominent in barren areas. The fungus provides water and a tight hold on a sandy or rocky substrate, while the alga provides food through its photosynthetic capacity.

**Parasitism** is a widespread nutritional strategy in which one member of the couple harms the other. The exploiting member is the **parasite**, the other member the **host**. Bacterial diseases involve a parasitic infestation by pathogenic microorganisms. Many species of flatworms and roundworms are parasites in vertebrates ranging from frogs to humans.

# *Index*